Recent Advances in Small-Angle Neutron Scattering

Recent Advances in Small-Angle Neutron Scattering

Editor

Sebastian Jaksch

MDPI • Basel • Beijing • Wuhan • Barcelona • Belgrade • Manchester • Tokyo • Cluj • Tianjin

Editor
Sebastian Jaksch
Jülich Centre for Neutron
Science at MLZ
Germany

Editorial Office
MDPI
St. Alban-Anlage 66
4052 Basel, Switzerland

This is a reprint of articles from the Special Issue published online in the open access journal *Applied Sciences* (ISSN 2076-3417) (available at: https://www.mdpi.com/journal/applsci/special_issues/small_angle_neutron_scattering).

For citation purposes, cite each article independently as indicated on the article page online and as indicated below:

LastName, A.A.; LastName, B.B.; LastName, C.C. Article Title. *Journal Name* **Year**, *Volume Number*, Page Range.

ISBN 978-3-0365-3729-0 (Hbk)
ISBN 978-3-0365-3730-6 (PDF)

© 2022 by the authors. Articles in this book are Open Access and distributed under the Creative Commons Attribution (CC BY) license, which allows users to download, copy and build upon published articles, as long as the author and publisher are properly credited, which ensures maximum dissemination and a wider impact of our publications.

The book as a whole is distributed by MDPI under the terms and conditions of the Creative Commons license CC BY-NC-ND.

Contents

About the Editor vii

Sebastian Jaksch
Recent Advances in Small-Angle Neutron Scattering
Reprinted from: *Appl. Sci.* **2022**, *12*, 90, doi:10.3390/app12010090 1

Sebastian Jaksch, Alexis Chennevière, Sylvain Désert, Tadeusz Kozielewski, Herbert Feilbach, Pascal Lavie, Romuald Hanslik, Achim Gussen, Stephan Butterweck, Ralf Engels, Henrich Frielinghaus, Stephan Förster and Peter Müller-Buschbaum
Technical Specification of the Small-Angle Neutron Scattering Instrument SKADI at the European Spallation Source
Reprinted from: *Appl. Sci.* **2021**, *11*, 3620, doi:10.3390/app11083620 5

Andreas Josef Schmid, Lars Wiehemeier, Sebastian Jaksch, Harald Schneider, Arno Hiess, Torsten Bögershausen, Tobias Widmann, Julija Reitenbach, Lucas P. Kreuzer, Matthias Kühnhammer, Oliver Löhmann, Georg Brandl, Henrich Frielinghaus, Peter Müller-Buschbaum, Regine von Klitzing and Thomas Hellweg
Flexible Sample Environments for the Investigation of Soft Matter at the European Spallation Source: Part I—The In Situ SANS/DLS Setup
Reprinted from: *Appl. Sci.* **2021**, *11*, 4089, doi:10.3390/app11094089 23

Tobias Widmann, Lucas P. Kreuzer, Matthias Kühnhammer, Andreas J. Schmid, Lars Wiehemeier, Sebastian Jaksch, Henrich Frielinghaus, Oliver Löhmann, Harald Schneider, Arno Hiess, Regine von Klitzing, Thomas Hellweg and Peter Müller-Buschbaum
Flexible Sample Environment for the Investigation of Soft Matter at the European Spallation Source: Part II—The GISANS Setup
Reprinted from: *Appl. Sci.* **2021**, *11*, 4036, doi:10.3390/app11094036 37

Matthias Kühnhammer, Tobias Widmann, Lucas P. Kreuzer, Andreas J. Schmid, Lars Wiehemeier, Henrich Frielinghaus, Sebastian Jaksch, Torsten Bögershausen, Paul Barron, Harald Schneider, Arno Hiess, Peter Müller-Buschbaum, Thomas Hellweg, Regine von Klitzing, and Oliver Löhmann
Flexible Sample Environments for the Investigation ofSoft Matter at the European Spallation Source: Part III—The Macroscopic Foam Cell
Reprinted from: *Appl. Sci.* **2021**, *11*, 5116, doi:10.3390/app11115116 53

Volker S. Urban, William T. Heller, John Katsaras and Wim Bras
Soft Matter Sample Environments for Time-Resolved Small Angle Neutron Scattering Experiments: A Review
Reprinted from: *Appl. Sci.* **2021**, *11*, 5566, doi:10.3390/app11125566 67

Xingxing Yao, Blake Avery, Miljko Bobrek, Lisa Debeer-Schmitt, Xiaosong Geng, Ray Gregory, Greg Guyotte, Mike Harrington, Steven Hartman, Lilin He, Luke Heroux, Kay Kasemir, Rob Knudson, James Kohl, Carl Lionberger, Kenneth Littrell, Matthew Pearson, Sai Venkatesh Pingali, Cody Pratt, Shuo Qian, Mariano Ruiz-Rodriguez, Vladislav Sedov, Gary Taufer, Volker Urban and Klemen Vodopivec
A Unified User-Friendly Instrument Control and Data Acquisition System for the ORNL SANS Instrument Suite
Reprinted from: *Appl. Sci.* **2021**, *11*, 1216, doi:10.3390/app11031216 99

Tetyana Kyrey, Marina Ganeva, Judith Witte, Artem Feoktystov, Stefan Wellert, Olaf Holderer
Grazing Incidence Small-Angle Neutron Scattering: Background Determination and Optimization for Soft Matter Samples
Reprinted from: *Appl. Sci.* **2021**, *11*, 3085, doi:10.3390/app11073085 113

Livia Balacescu, Georg Brandl, Fumitoshi Kaneko, Tobias Erich Schrader and Aurel Radulescu
Light Scattering and Absorption Complementarities to Neutron Scattering: In Situ FTIR and DLS Techniques at the High-Intensity and Extended Q-Range SANS Diffractometer KWS-2
Reprinted from: *Appl. Sci.* **2021**, *11*, 5135, doi:10.3390/app11115135 123

Ezzeldin Metwalli, Klaus Götz, Tobias Zech, Christian Bär, Isabel Schuldes, Anne Martel, Lionel Porcar and Tobias Unruh
Simultaneous SAXS/SANS Method at D22 of ILL: Instrument Upgrade
Reprinted from: *Appl. Sci.* **2021**, *11*, 5925, doi:10.3390/app11135925 141

Inna A. Belyaeva, Jürgen Klepp, Hartmut Lemmel and Mikhail Shamonin
Feasibility of Probing the Filler Restructuring in Magnetoactive Elastomers by Ultra-Small-Angle Neutron Scattering
Reprinted from: *Appl. Sci.* **2021**, *11*, 4470, doi:10.3390/app11104470 153

Daniël N. ten Napel, Janne-Mieke Meijer and Andrei V. Petukhov
The Analysis of Periodic Order in Monolayers of Colloidal Superballs
Reprinted from: *Appl. Sci.* **2021**, *11*, 5117, doi:10.3390/app11115117 163

About the Editor

Sebastian Jaksch

Sebastian Jaksch obtained his PhD working on thermoresponsive polymers at the Technical University of Munich, Germany and continued his work at the Jülich Centre for Neutron Science, based at the FRM 2 research reactor, in nearby Garching. Here, he expanded his research to include biological systems, using his previous experience with polymers in solution to investigate polymeric drug delivery systems. In addition to his work on soft-matter he also strived to improve the experimental possibilities for neutron scattering on such systems. To this end, he developed the Small-Angle Neutron Scattering (SANS) instrument SKADI for the European Spallation Source in Lund, Sweden as well as a corresponding neutron detector, SoNDe, for high-flux applications. He has also helped to establish the grazing incidence neutron spin-echo spectroscopy method for investigating supported lipid membranes. During the COVID-19 pandemic he started working with virus-derived materials to investigate viral fusion.

Editorial

Recent Advances in Small-Angle Neutron Scattering

Sebastian Jaksch

Jülich Centre for Neutron Science JCNS at Heinz Maier-Leibnitz Zentrum (MLZ),
Forschungszentrum Jülich GmbH, Lichtenbergstr. 1, 85748 Garching, Germany; s.jaksch@fz-juelich.de

Citation: Jaksch, S. Recent Advances in Small-Angle Neutron Scattering. *Appl. Sci.* **2022**, *12*, 90. https://doi.org/10.3390/app12010090

Received: 15 December 2021
Accepted: 17 December 2021
Published: 22 December 2021

Publisher's Note: MDPI stays neutral with regard to jurisdictional claims in published maps and institutional affiliations.

Copyright: © 2021 by the author. Licensee MDPI, Basel, Switzerland. This article is an open access article distributed under the terms and conditions of the Creative Commons Attribution (CC BY) license (https://creativecommons.org/licenses/by/4.0/).

Small-angle scattering, and its neutron expression small-angle neutron scattering (SANS), has developed into an invaluable tool for the investigation of microscopic and mesoscopic structures in recent decades. This Special Issue aims to provide an overview of the recent experimental and technological advances of this technique. The knowledge of advances in SANS, being ubiquitous in physical sciences, therefore promises to help a wide range of researchers in their respective fields, whether that is physical chemistry, biophysics, pharmaceutic and medical studies or the studies of magnetic phenomena.

This overview comprises the description of new instrumentation, sample environments and instrument control and data handling, over new experimental applications and their combination with complementary experimental techniques.

The instrumentation descriptions deliver an in-depth understanding of the status of the technical capabilities of SANS in current instrumentation as well as the sample environment. With this knowledge, experimenters can now better gauge the technical feasibility of their experiments, which is both important for young researchers who are new to the field, but also to more experienced researchers who may want to revisit experiments which were not feasible at an earlier point in time. Both approaches promise new scientific insights and allow using SANS to capacity. Additionally, scientists at facilities which provide SANS instruments may use some of the possibilities shown for sample environments to widen their experimental range.

In "Technical Specification of the Small-Angle Neutron Scattering Instrument SKADI at the European Spallation Source (ESS)" [1], the capacity and capability of a state-of-the-art SANS instrument is presented, which is currently constructed at the ESS. Some of the features described are not fundamentally based on the higher brilliance of the source as such, so those can in principle also be realized at existing facilities.

Closely linked to that are three papers on flexible sample environments at the ESS [2–4]. While they are based on the technical concept of SKADI, the ideas should be universally applicable and support original ideas by other researchers.

Two other papers are primarily linked to the facilities of the Oak Ridge National Laboratory (ORNL), and give a review of the soft matter sample environments for SANS available there [5] and present the newest instrumental control system [6]. While this is particularly important to potential users of the ORNL facilities, those ideas can again be used as inspiration.

The last technical paper in this issue focuses on the impact of SANS background in the case of grazing incidence SANS (GISANS) [7]. GISANS has been developed into a method in its own right over the last decades; however, it still remains closely linked to SANS, both scientifically and technologically. Considering the mutual impacts and differences allows for a more complete understanding of the measured data.

The same is true for the description of complementary techniques. Additionally, here, inspiration for new experiments both on new and already otherwise-investigated samples can be taken from the manuscripts published in this issue.

Showing two complementary in situ techniques for SANS "Light Scattering and Absorption Complementarities to Neutron Scattering: In Situ FTIR and DLS Techniques at the High-Intensity and Extended Q-Range SANS Diffractometer KWS-2" [8] describes a setup where two methods are combined with SANS, both of which deliver data which are otherwise unobtainable in SANS yet help with a comprehensive data evaluation.

The description of simultaneous SANS and small-angle X-ray scattering (SAXS) in "Simultaneous SAXS SANS Method at D22 of ILL Instrument Upgrade" [9] shows an additional example of complementary measurements and the achievable benefits of simultaneous in situ measurements. Even though both methods are small-angle scattering and thus governed by similar principles, the combination of two independent contrast systems can illuminate features which would otherwise be hidden, especially in the case of soft matter, where the contrast differences for hydrogenated and deuterated samples are very pronounced.

As a final, but important, component of this issue, new ingenious experiments and their evaluations are featured. Those state-of-the-art applications of the SANS technology may likewise broaden the view of SANS usage for new experiments.

In "Feasibility of Probing the Filler Restructuring in Magnetoactive Elastomers by Ultra-Small-Angle Neutron Scattering" [10], SANS as a method to investigate material properties on the micrometer scale in bulk is described. Here, future research approaches for such systems, such as polarized neutron scattering and scattering at low temperatures, are also discussed.

While the manuscript "The Analysis of Periodic Order in Monolayers of Colloidal Superballs" [11] strictly only uses SAXS data, the method applied for the analysis is fully applicable for SANS experiments. Since such systems attract the interest of a large community and also show a high degree of technical applicability, this analytical method is also of interest for researchers focused on SANS measurements.

While for such a wide field as SANS there cannot be any claim for completeness, this Special Issue is a good representation of the technologies available, newly developed and newly applied in the field of SANS. With each article giving an in-depth description of the particular described feature, technique or approach, it can be hoped the articles collected in this issue may help the readers to conceive both new and improved ideas about the application of the SANS technique in their respective fields.

Funding: This research received no external funding.

Institutional Review Board Statement: Not applicable.

Informed Consent Statement: Not applicable.

Data Availability Statement: Not applicable.

Conflicts of Interest: The authors declare no conflict of interest.

References

1. Jaksch, S.; Chennevière, A.; Désert, S.; Kozielewski, T.; Feilbach, H.; Lavie, P.; Hanslik, R.; Gussen, A.; Butterweck, S.; Engels, R.; et al. Technical Specification of the Small-Angle Neutron Scattering Instrument SKADI at the European Spallation Source. *Appl. Sci.* **2021**, *11*, 3620. [CrossRef]
2. Kühnhammer, M.; Widmann, T.; Kreuzer, L.P.; Schmid, A.J.; Wiehemeier, L.; Frielinghaus, H.; Jaksch, S.; Bögershausen, T.; Barron, P.; Schneider, H.; et al. Flexible Sample Environments for the Investigation of Soft Matter at the European Spallation Source: Part III—The Macroscopic Foam Cell. *Appl. Sci.* **2021**, *11*, 5116. [CrossRef]
3. Schmid, A.J.; Wiehemeier, L.; Jaksch, S.; Schneider, H.; Hiess, A.; Bögershausen, T.; Widmann, T.; Reitenbach, J.; Kreuzer, L.P.; Kühnhammer, M.; et al. Flexible Sample Environments for the Investigation of Soft Matter at the European Spallation Source: Part I—The In Situ SANS/DLS Setup. *Appl. Sci.* **2021**, *11*, 4089. [CrossRef]
4. Widmann, T.; Kreuzer, L.P.; Kühnhammer, M.; Schmid, A.J.; Wiehemeier, L.; Jaksch, S.; Frielinghaus, H.; Löhmann, O.; Schneider, H.; Hiess, A.; et al. Flexible Sample Environment for the Investigation of Soft Matter at the European Spallation Source: Part II—The GISANS Setup. *Appl. Sci.* **2021**, *11*, 4036. [CrossRef]
5. Urban, V.S.; Heller, W.T.; Katsaras, J.; Bras, W. Soft Matter Sample Environments for Time-Resolved Small Angle Neutron Scattering Experiments: A Review. *Appl. Sci.* **2021**, *11*, 5566. [CrossRef]
6. Yao, X.; Avery, B.; Bobrek, M.; Debeer-Schmitt, L.; Geng, X.; Gregory, R.; Guyotte, G.; Harrington, M.; Hartman, S.; He, L.; et al. A Unified User-Friendly Instrument Control and Data Acquisition System for the ORNL SANS Instrument Suite. *Appl. Sci.* **2021**, *11*, 1216. [CrossRef]

7. Kyrey, T.; Ganeva, M.; Witte, J.; Feoktystov, A.; Wellert, S.; Holderer, O. Grazing Incidence Small-Angle Neutron Scattering: Background Determination and Optimization for Soft Matter Samples. *Appl. Sci.* **2021**, *11*, 3085. [CrossRef]
8. Balacescu, L.; Brandl, G.; Kaneko, F.; Schrader, T.E.; Radulescu, A. Light Scattering and Absorption Complementarities to Neutron Scattering: In Situ FTIR and DLS Techniques at the High-Intensity and Extended Q-Range SANS Diffractometer KWS-2. *Appl. Sci.* **2021**, *11*, 5135. [CrossRef]
9. Metwalli, E.; Götz, K.; Zech, T.; Bär, C.; Schuldes, I.; Martel, A.; Porcar, L.; Unruh, T. Simultaneous SAXS/SANS Method at D22 of ILL: Instrument Upgrade. *Appl. Sci.* **2021**, *11*, 5925. [CrossRef]
10. Belyaeva, I.A.; Klepp, J.; Lemmel, H.; Shamonin, M. Feasibility of Probing the Filler Restructuring in Magnetoactive Elastomers by Ultra-Small-Angle Neutron Scattering. *Appl. Sci.* **2021**, *11*, 4470. [CrossRef]
11. Napel, D.N.T.; Meijer, J.-M.; Petukhov, A.V. The Analysis of Periodic Order in Monolayers of Colloidal Superballs. *Appl. Sci.* **2021**, *11*, 5117. [CrossRef]

Article

Technical Specification of the Small-Angle Neutron Scattering Instrument SKADI at the European Spallation Source

Sebastian Jaksch [1,2,*], Alexis Chennevière [3], Sylvain Désert [3], Tadeusz Kozielewski [4], Herbert Feilbach [4], Pascal Lavie [3], Romuald Hanslik [5], Achim Gussen [5], Stephan Butterweck [5], Ralf Engels [4], Henrich Frielinghaus [1], Stephan Förster [1,4] and Peter Müller-Buschbaum [2,6]

1. Forschungszentrum Jülich GmbH, JCNS at Heinz Maier-Leibnitz Zentrum, Lichtenberstraße 1, 85748 Garching, Germany; h.frielinghaus@fz-juelich.de (H.F.); s.foerster@fz-juelich.de (S.F.)
2. Heinz Maier-Leibnitz Zentrum (MLZ), Technische Universität München, Lichtenberstraße 1, 85748 Garching, Germany; muellerb@ph.tum.de
3. LLB, CEA, CNRS, Université Paris-Saclay, 91191 Gif-sur-Yvette, France; alexis.chenneviere@cea.fr (A.C.); sylvain.desert@cea.fr (S.D.); pascal.lavie@cea.fr (P.L.)
4. Forschungszentrum Jülich GmbH, Jülich Centre for Neutron Science (JCNS), 52425 Jülich, Germany; t.kozielewski@fz-juelich.de (T.K.); h.feilbach@fz-juelich.de (H.F.); r.engels@fz-juelich.de (R.E.)
5. Forschungszentrum Jülich GmbH, Zentralinstitut für Engineering, Elektronik und Analytik—Engineering und Technologie (ZEA-1), 52425 Jülich, Germany; r.hanslik@fz-juelich.de (R.H.); a.gussen@fz-juelich.de (A.G.); s.butterweck@fz-juelich.de (S.B.)
6. Lehrstuhl für Funktionelle Materialien, Physik-Department, Technische Universität München, James-Franck-Str. 1, 85748 Garching, Germany
* Correspondence: s.jaksch@fz-juelich.de

Abstract: Small-K Advanced DIffractometer (SKADI is a Small-Angle Neutron Scattering (SANS) instrument to be constructed at the European Spallation Source (ESS). SANS instruments allow investigations of the structure of materials in the size regime between Angstroms up to micrometers. As very versatile instruments, they usually cater to the scientific needs of communities, such as chemists, biologists, and physicists, ranging from material and food sciences to archeology. They can offer analysis of the micro- and mesoscopic structure of the samples, as well as an analysis of the spin states in the samples, for example, for magnetic samples. SKADI, as a broad range instrument, thus offers features, such as an extremely flexible space for the sample environment, to accommodate a wide range of experiments, high-flux, and optimized detector-collimation system to allow for an excellent resolution of the sample structure, short measurement times to be able to record the internal kinetics during a transition in the sample, as well as polarized neutron scattering. In this manuscript, we describe the final design for the construction of SKADI. All of the features and capabilities presented here are projected to be included into the final instrument when going into operation phase.

Keywords: SANS; small-angle neutron scattering; SKADI; ESS; European Spallation Source

1. Introduction

The Small-K Advanced DIffractometer (SKADI, for the naming we exploited the correspondence between the reciprocal space vector $\vec{K}$ and momentum transfer vector $\vec{Q}$) is a joint in-kind project of French (LLB) and German (FZ Jülich) partners to deliver a Small-Angle Neutron Scattering (SANS) instrument to the European Spallation Source (ESS) [1,2]. The originally proposed instrument is described in a previous publication [3]. This manuscript details the further developments of SKADI, highlighting the final changes at the beginning of the construction phase. In addition, further practical requirements for performing experiments, such as sample area and environment, will be considered. Here, we want to highlight the extensive collaboration with user groups in order to develop state-of-the art sample environments [4–6].

Small-Angle Neutron Scattering (SANS) instruments scatter neutrons from samples at angles usually below 0.3 rad and above some tenth of a mrad. Combined with cold neutrons at wavelengths above ~3 Å this results in an accessible inverse space of $1\,\text{Å}^{-1} \geq Q \geq$ a few 10^{-4}Å^{-1} and a corresponding size range in real space in the sample between several Angstroms and a micrometer. The spatial resolution at the detector, the wavelength resolution of the detected neutrons, the brilliance transfer of useful neutrons to the sample, the simultaneously covered solid angle with the detector, as well as the general suitability for a wide range of experiments are important parameters during the design of such an instrument.

While the spatial resolution of the detector itself is given by the employed detector technology, the positioning of the detector can be used to adjust the solid angle coverage of a pixel or the readout mode [7] to achieve the desired resolution. For most experimental settings, the detector resolution of SKADI is better than needed in order to avoid an additional contribution to the overall resolution. Thus, for example, for a central detector setting (sample-detector distance equals the collimation length), an ideal mapping of the sample aperture would lead to a pixel size identical to that of the generally used sample aperture ($10 \times 10\,\text{mm}^2$), while the chosen pixel resolution is $6 \times 6\,\text{mm}^2$. On the other hand, for the shorter detector position of the front detector, this still improves the resolution.

The wavelength resolution of the instrument is mainly determined by the rejection of unwanted neutrons in a selector or, as is the case with SKADI, by the time discrimination of arriving neutron pulses with a spread of kinetic energy and a time-of-flight readout of the detector.

In order to transfer the highest amount of neutrons per second, solid angle, angular divergence, and time, i.e., brilliance, to the sample the complete instrument has to be considered, since any component that is influencing the flow of neutrons can also absorb or deflect them. Special focus, in this case, has to be given to the extraction of neutrons from the moderator in order to maximize the brilliance transfer while avoiding a direct line of sight between sample and moderator to minimize unwanted background, for example, from γ-radiation.

The general feasibility for different experiments is again a holistic issue concerning all components of the instrument. The instrument needs to facilitate large, unconventional, or custom made sample environments, delivering the correct choice of resolution for the experiment at hand, while performing it in a timely manner. The control software needs to allow to adjust the experimentally relevant parameters, while allowing a good overview of everything that is happening at the same time and not distract the user from the experiment at hand. For SKADI, in fact, the whole instrument suite of ESS, NICOS [8] as a control software was chosen, and the appropriate requirements are being implemented. The data will be recorded in event-mode, i.e., with a time-stamp for each single neutron event on the detector. This will allow for cross-referencing additional settings of motor positions or sample environments to a specific neutron event. Finally, the data that will be provided to the user need to be in a format that facilitates the work flow of data analysis.

The description will treat the components following the path of the neutrons from the moderator to the detector in order to give a full overview of SKADI in a comprehensive manner. All of the simulations shown here have been carried out using the McStas framework [9].

2. Overview

Figure 1 and Table 1 give an overview of the SKADI instrument. The different components are described in sequence of the neutron flux from the neutron extraction up until the detector. There is a dedicated section for each of the described components.

Table 1. Tabulated instrumental parameters for Small-K Advanced DIffractometer (SKADI).

	Quick Facts
Spectrum	Cold
Overall Length of Instrument	58 m
Q-range	a few $10^{-4} - 1 \text{Å}^{-1}$
Spin-Polarization/-Analysis	yes / planned
Wavelength band	5 (standard mode) or 10 Å (pulse skipping mode)
Wavelength range	3–21 Å
Momentum resolution	2–7% (standard mode) or 1–7% (pulse skipping mode)
Flux at sample	8.8×10^8 neutrons s^{-1} cm^{-2} @ 5 MW accelerator power and 4 m collimation

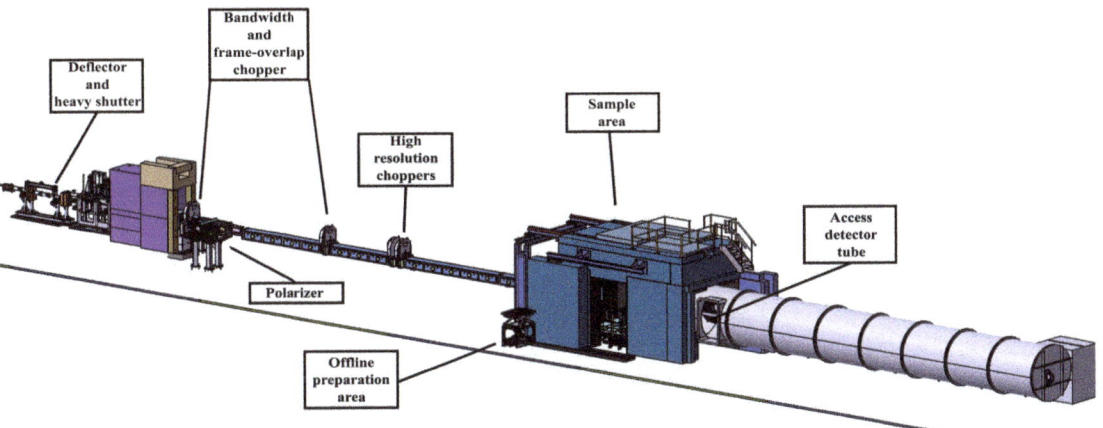

Figure 1. SKADI overview image. The scale bar is in sections of 10 m. From the moderator to the extraction there is a 2 m free flight path up to the neutron extraction. Inside the bunker, there is the neutron extraction of a double deflector type (see Section 2.1). This section also includes the heavy shutter, which vertically replaces the last section of neutron guide with an absorption block. The bunker wall is located between 11 and 15.5 m. Directly outside the bunker wall is the polarizer housing with an integrated chopper. Inside the collimation (from the bunker wall to the sample area), there is a other chopper disk at 22.85 m and the infrastructure for including two high resolution choppers further downstream. The collimation feeds into the sample area with the complete infrastructure for mounting the different setups for sample environments. The detector tube is directly attached to the sample area. The shielding is removed for better visibility.

2.1. Neutron Extraction

The neutron extraction is responsible for avoiding a direct line of sight between the moderator and the sample while transporting the highest brilliance possible. The two main considerations making this out of line of sight requirement necessary are radiation protection and signal to noise ratio. In the moderator area there is a high flux of particles other than neutrons, from the spallation process, and virtually any particle possible created by bremsstrahlung for which the accelerator provides enough energy. In addition to that, there is approximately one γ per neutron and high energy γ radiation may produce photo neutrons. Indeed, the signal to noise ratio with direct view of the moderator would be poor due to those additional particles. Existing detector technologies can hardly discriminate between that many different particles at such high flux rates. Even neglecting the necessity of having a clean spectrum at the sample and detector position, most of these particles would require higher shielding efforts when compared to cold or thermal neutrons. This is even more the case for an instrument like SKADI, which only requires the cold flux from the moderator at very low divergencies. Furthermore, reducing the total flux of particles, the requirements on shielding are considerably more moderate.

For cold neutrons, there are, in general, three approaches to that problem: (1) curved neutron guides (benders), (2) doglegs, and (3) deflectors. In the case of the curved neutron guides, channeled benders should also be considered. In order to transport the highest brilliance, and not merely the highest flux with highly diverging neutrons, it is a good approach to transport the ideal image of a fraction of the moderator equal to the cross section of the last defining aperture before the sample aperture. Higher divergent neutrons could possibly be transported, however would only be collimated away in the later parts of the instrument and increase shielding effort, while all of the neutrons as described before, can be used for the scattering experiment. In addition to that function, fast neutrons should be removed in the neutron extraction, which is usually happening as a by-product, since fast neutrons will not be affected by most of the optical components and, thus, propagate on a straight path.

Thus, the first issue is finding the optimal view of the moderator in terms of flux and spectrum. Figure 2 shows an image of integrated intensity of useful neutrons at the deflector exit vs. horizontal angle and horizontal shift of the neutron extraction system along the inside of the monolith. A diagonal line of optimal flux is found because the moderator has a preferred direction in which the neutrons are transmitted. Pairing this with the geometrical restrictions of the monolith insert a preferred window of installation is found. Using this, we chose to position the optical axis of SKADI at the coordinates of $\alpha_{hor} = -0.5°$ and $x_{offset} = 7$ mm. That is fairly central in the possible window and close to the central region of the high intensity ridge.

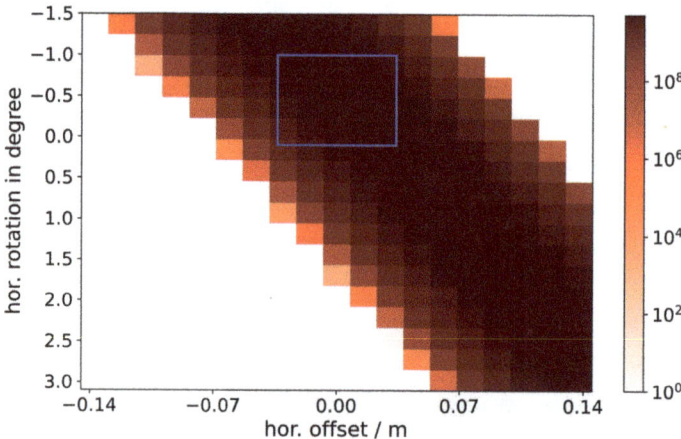

Figure 2. The intensity map for the moderator view relating horizontal offset of the extraction system along the inside of the moderator and rotation. A window for physically possible settings is outlined in blue. The ridge of highest intensity corresponds to the ratio between offset and rotation to keep looking at the location of highest flux inside the moderator. The inclination of that ridge is angle = 27.3 × offset.

The deflector, dogleg, and bender were examined in order to evaluate the optimum transport mechanism from the monolith to the bunker wall. Restrictions on the materials were the exclusive use of metal substrates inside the bunker and a maximum m-value of m = 4 to prevent too strong degradation due to the high radiation field inside the bunker [10]. The available length inside the bunker is 9 m, including 3.5 m of the monolith insert. The minimum distance to bridge in vertical direction was 20 cm, as this is the width of the anticipated hadronic shower coming from the target and the moderator that should not propagate along the neutron guide to the sample. By vertically shifting the neutron guide by the specified 20 cm or more, this background can be harmlessly absorbed in the bunker wall. This vertical displacement also has the advantage of keeping the instrument straight within its assigned angular wedge and avoid interference with neighbouring instruments.

The deflector neutron extraction is the finally chosen solution for the neutron extraction system. The most striking reason for that is given by the performance shown in Figure 3. Looking at the spectra and the detailed data of the deflector explains this behavior. As detailed above, one of the primary objectives of the extraction is to block the direct line of sight. This is most easily achieved by inclining the neutron guide or, as in this case, the upper section of the neutron guide until the full cross section of the neutron guide is covered. For SKADI, this means that 3 cm needs to be covered. The restriction of m = 4 and the lowest wavelength of neutrons to be transported of $\lambda = 3$ Å leads to the following onsiderations:

The critical scattering angle of m = 4 is $Q_{crit} = 0.087$ Å^{-1}. Using this to find the angle of inclination for the upper surface of the neutron guide, we need to use

$$Q_{crit} = \frac{4\pi \sin \theta}{\lambda} \rightarrow \theta = \arcsin \frac{Q_{crit} \lambda}{4\pi}, \tag{1}$$

which yields $\theta = 0.88°$, using $\lambda = 2$ Å. We chose the lower value for λ for this calculation in order to increase the window of accepted neutrons. Finally, we chose $\theta = 0.9°$ as the engineering requirement. This shifts the accepted wavelengths to higher values, which is not a drawback as the targeted wavelength band starts at $\lambda = 3$Å. Because the drop in intensity at the critical edge is centered around $Q_{crit} = 0.087$Å^{-1} and being in the area of the highest flux, there may still be significant contributions at higher Q-values. Additionally, this gives a slightly relaxed engineering margin, as now the complete cross section of the neutron guide can be covered by an inclined upper neutron mirror in the guide within 2 m using 2 m $\times \sin 0.9° = 3.14$ cm. Those relaxations of the exact requirement, the slightly sharper inclined mirror as well as the larger coverage of the neutron guide cross section benefit the function of rejecting low wavelengths efficiently and ensuring that the line of sight is covered. The same setup that is rotated by 180° is used at the exit end of the neutron extraction, since the neutrons have to be reflected into the horizontal plane again. This means the middle section is 5 m long and the inclination of that middle section should be twice the inclination of the entrance and exit mirrors in order to transport the neutrons with as few reflections as possible. Thus, the central section is a 5 m neutron guide, being inclined by 1.8°. Figure 4 depicts a drawing of the cross section. The entry and exit part mainly function as an inclined mirror to deflect the desired neutrons by the necessary amount. This functions much like a periscope. The total deflection of the beam that is achieved by this setup is 22 cm, which fulfills the requirement of more than 20 cm. This double reflection should reduce the production of secondary particles in the second footprint of the beam in the supermirror to suppress background. Moreover, the primary hadronic particle shower of the target onto the bunker wall is calculated to be approximately 20×20 cm^2, i.e., 10 cm in vertical direction from the center of the beam. Offsetting by 22 cm should also reliably suppress that source of background.

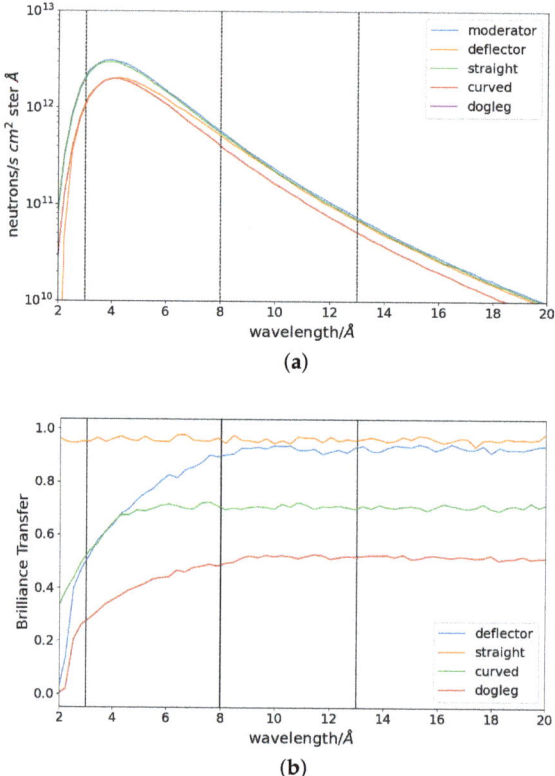

Figure 3. (**a**) Spectra and (**b**) brilliance transfer curves for different solutions of neutron extraction in the case of SKADI. Brilliance transfer was considered for the wavelength shown at divergencies up to 0.0286°, which corresponds to a 1 cm aperture at a collimation length of 20 m. Because SKADI is a small-angle instrument, performance at the lowest angles is important.

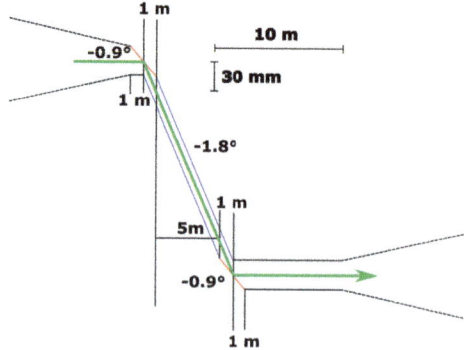

Figure 4. Sketch drawing of the deflector setup. This shows a vertical section through the deflector. Red lines designate m = 4 supermirrors, inclined by 0.9°, blue lines indicate m = 4 supermirrors inclined by 1.8°. The sides are coated with m = 1 supermirrors, since no vertical displacement of the beam center in necessary. The dashed lines at the beginning and end outline the transported divergence for the shortest collimation with the widest opening (3 cm beam cross section over 4 m collimation length). The central neutron beam is indicated in green.

2.2. Chopper

Details regarding the chopper setup of SKADI have already been published, including the analytical calculations used to develop a balanced chopper setup [11]. SKADI will provide the full infrastructure for four choppers, only equipping the first two positions (bandwidth and frame-overlap). High resolution choppers thus can be added at a later time. This results in chopper positions of 15.5 and 22.85 m (for 7 and 14 Hz) with additional chopper pits at 26.3 and 26.6 m. The choppers are optimized for a transmission of neutron wavelengths between 3 and 8 Å and offer the option for pulse skipping mode to achieve a wider wavelength band of 3 to 13 Å.

It should be noted that the simulation always assumes perfectly absorbing chopper disks. Our workshop can reliably produce chopper disks absorbing better than 10^{-7} that are averaged over the full shaded area.

In order to elucidate the impact of high resolution choppers, the different achievable resolutions for simulated δ-peaks are shown in Figure 5. Here, we compare the resolution improvement by either the additional choppers or a reduction of the slit size while only using two choppers. While the wavelength resolution $\Delta\lambda/\lambda$ can be considerably improved by using four instead of two choppers, this does not fully feed into a much improved Q-resolution $\sigma Q/Q$, as in the pertinent Q-range the aperture at sample position and the pixel size of the detector play a dominant role. However, looking at the simulated scattering from a sample creating exact δ-peaks, it can be seen that the background is lowered by approximately 10% when comparing the smaller apertures with the 1 cm apertures combined with high resolution choppers. Meanwhile, the peak width is nearly unchanged. This means a better signal-to-noise ratio can be achieved. Here, it also should be noted that the absolute count rate of contributing neutrons (neutrons that are scattered within a peak) is bigger by a factor of 2 in the case of the four chopper solution.

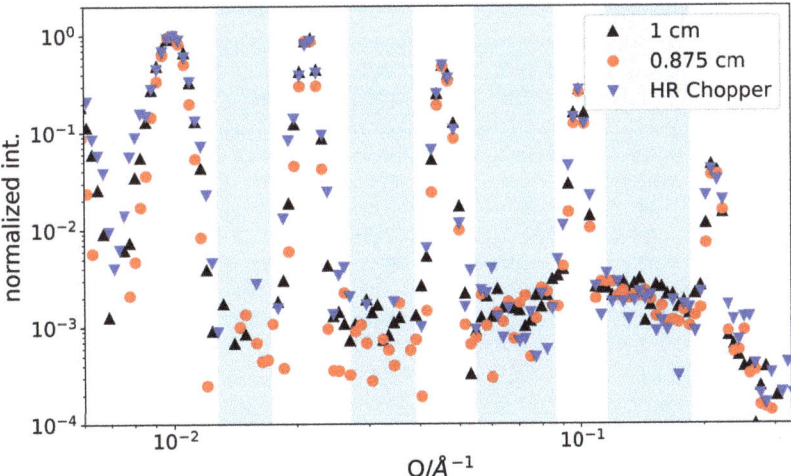

Figure 5. A set of simulated δ-peaks for different instrument configurations. While there is no major impact on the resolution for the additional choppers, the background is considerably lowered. Additionally, the absolute number of neutrons transmitted in the useful Q-band is higher by a factor of 2 for the chopper setup. Simulations where performed for 20 m collimation and sample apertures of 1 cm (▲), 0.875 cm (●), and with 1 cm sample aperture using the high-resolution choppers (▼). For better visibility, the lower cutoff is at a normalized intensity of 1×10^{-4}. In order to ascertain the respective background intensities, the intensities have been summed over the areas that are shaded blue, which results in 0.074 for the 1 cm slit opening, 0.062 for the 0.875 cm slit opening, and 0.056 for the high resolution chopper setup.

Because the choppers also should inhibit frame overlap between the different pulses, it needs to be considered that this function is still adequately performed by a set of two choppers. Simulations showed that at 40 Å an additional signal is visible with a relative intensity of 10^{-6} compared to the 3–8 Å wavelength band. For initial operation, this seems to be appropriate, at a later time a frame overlap mirror may improve this signal ratio, even without the use of additional choppers. Here, it also should be noted that, in addition the 40 Å neutrons will be attenuated far stronger by any window material. Additionally, in the case of a sufficiently long collimation, they will drop out of the collimation and be absorbed (∼0.4 mm drop per meter flight-path for a 40 Å neutron).

2.3. Collimation

The collimation allows for four distinct collimation lengths of 20, 14, 8, and 4 m. All of the collimation slits can be set between the full beam width of $3 \times 3\,\text{cm}^2$ and fully closed with independently moving blades from top, bottom, and both sides. Changing the collimation length means that at the chosen collimation length an aperture is inserted into the beam. All the neutron guides from the moderator to the aperture are left transporting the neutrons to the aperture. All of the neutron guides afterwards are removed from the beam. In the case of SKADI, the removed guides and slits are replaced by $35 \times 35\,\text{mm}^2$ passive blinds made from boron carbide [12]. Those remove stray neutron propagating parallel to the main beam and reduce the background. The neutron guides will be placed on sledges to be moved in and out of the beam while maintaining the alignment.

Additionally to the classical pin-hole setup, a very small-angle neutron scattering (VSANS) setup with fixed slit-collimators will allow for accessing Q-values down to several times $10^{-5}\,\text{Å}^{-1}$ [13].

2.4. VSANS

With a very small-angle scattering (VSANS) setup, SKADI will be able to reach Q-values down to several times $Q = 10^{-5}\,\text{Å}^{-1}$. This VSANS section will be inserted behind the last collimation blind at 4 m collimation length. The collimator will be 2 m long and made out of six radial collimators of 33 cm length each. They will be precisely aligned on an optical bench. The slits will be 3 cm high, 100 µm thick, and coated with Gadolinium. There will be 50 lateral slits to cover the 3 cm wide beam. The VSANS radial collimator is designed to produce a beam of approximately 3 mm, approximatively the size of the pixels of the central detector, at 20 m sample-detector distance [14,15].

This means that, in the final setup, there will be a slit-convoluted signal, of 50 parallel slits, on the detector. Treatment of those slit geometries has long been a standard in the literature [16,17], so data treatment should be straight forward. Tests of this setup have already been performed at LLB, CEA Saclay, Gif-Sur-Yvette. In these tests, only a single collimator slit and a point-like beamspot (30 mm × 96 mm) was used. Figure 6 shows the corresponding scattering data. From the data shown, it is also apparent that, at SKADI, a higher detector resolution and smaller usable slit size (at still acceptable flux) will directly lead to smaller achievable Q-values.

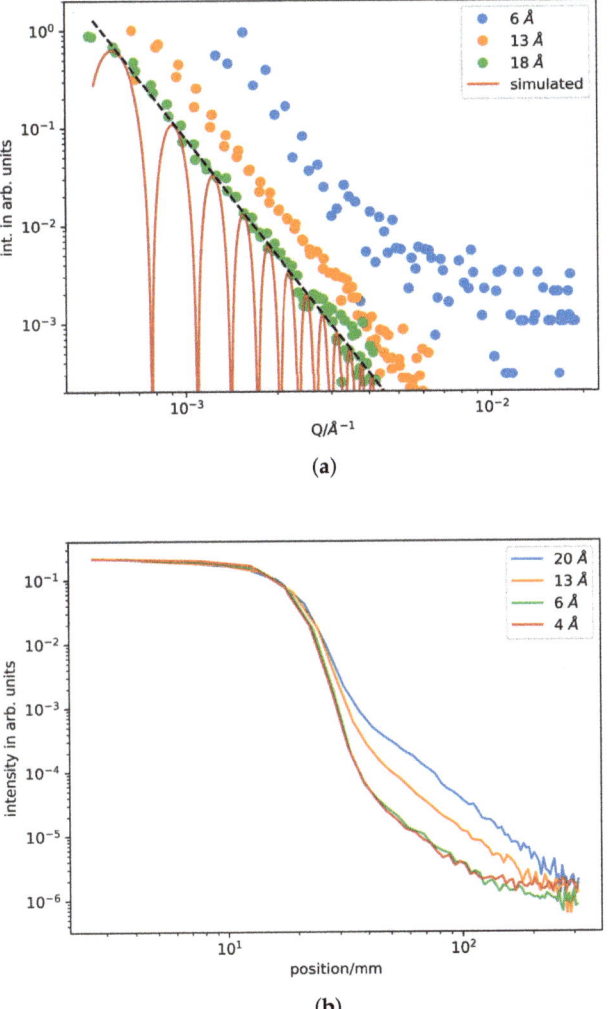

Figure 6. (**a**) Scattered intensity of 1 μm latex particles in D_2O for several wavelengths (•) and from simulation (solid line); (**b**) beam profile on detector for the indicated wavelengths. Both datasets were recorded on the PA20 instrument at LLB, CEA Saclay. When comparing the simulated and recorded scattering of the latex particles, it becomes clear that the minima of such particles could not be resolved by the given instrument resolution. This results from the pixel size of the detector being 5 mm × 5 mm and the beam spot size of about 30 mm in the closest direction. However, the slope of the maxima is reproduced well (dashed line reproduces $I \propto Q^{-4}$ behavior). The difference in intensities is due to the spectrum available at the instrument. When considering the beam spot size, it becomes apparent that, at higher wavelength, the Gd of the coating starts to contribute more strongly to be scattering. This leaves an optimization process of the smallest usable Q-values to hot commissioning. Probably strongly scattering samples can achieve a lower Q-value, since the additional contribution due to the radial collimator is limited.

2.5. Polarizer and Guide Field

In order to provide polarized neutrons with an optimal degree of polarization over the complete wavelength band SKADI will feature two polarizers optimized for the low (λ = 3–8 Å) and high (λ = 8–13 Å) wavelength band. This is achieved by two cavities with a single slanted supermirror that provides the polarization. Figure 7 presents the respective polarization and transmission. One cavity is a double coated Fe-Si supermirror with m = 2.5 in V-geometry using the corresponding McStas component [9]. Here, one additional polarizer position is left unoccupied for future upgrades.

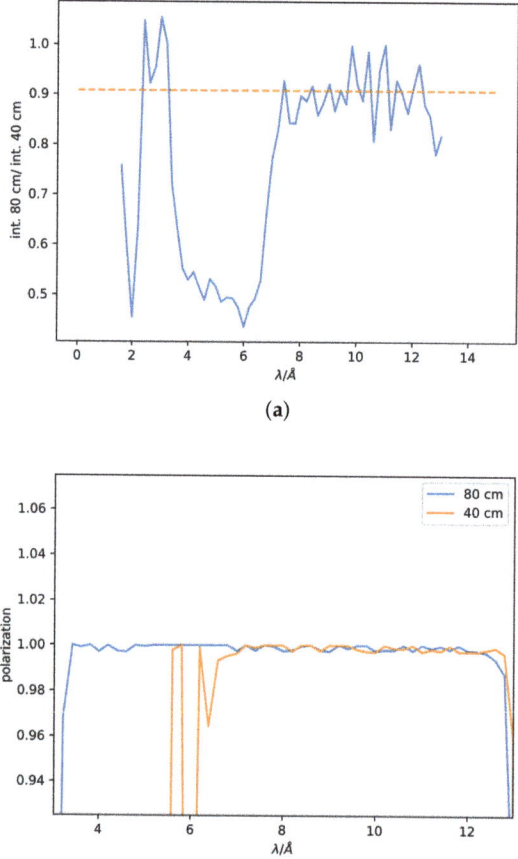

Figure 7. (a) Relative transmission of the two polarizer cavities, (b) polarization at sample position for the respective polarizer cavities. Simulation was conducted for a 20 m collimation with 1 cm × 1 cm beam size and in pulse skipping mode (i.e., 3–13 Å wavelength band). In order to optimize both, transmission and polarization, the longer cavity is used for the short wavelength band, while the short cavity is used for the long wavelength band. Thus, the poor polarization at low wavelength is avoided for the shorter cavity, and still the higher transmission in the high wavelength band can be used. The vertical line is a fit of the relative intensity between the two cavities for the wavelength between 8 and 13 Å and it indicates a transmission of 90.8% for the 80 cm cavity when compared to the 40 cm cavity. The decreasing polarization at the boundaries of the used wavelength band are binning effects due to zero intensity outside the shown wavelength band.

In order to maintain the polarization throughout the collimation, it is necessary to keep a homogeneous magnetic field after the polarizer until the sample position. While the field strength may vary, the direction needs to be consistent and should never flip. Calculations for permanent magnets that are aligned along the collimation show that a field strength between 5 and 13 mT can be maintained throughout the collimation, including all components, such as choppers. The chosen field direction is perpendicular to the beam direction. To also allow for different spin orientations, a spin flip is realized by a radio-frequency spin flipper directly after the polarizer [12]. The foreseen probability for a successful spin flip is $\geq$99%.

2.6. Sample Area

The sample area, as such, is the main area where the experimenter will feel the strongest impact of the design in terms of usability. A kinematic mounting system will be installed in order to allow for an effective and speedy setup of the different experiments to be performed at SKADI. This kinematic mounting system will allow an off-instrument preparation of the experimental setup and several user groups have already developed corresponding sample environments [4–6]. This setup is designed to take up a load of up to 990 kg on the area of a EUR-pallet. It is limited to $\leq$1000 kg to fulfill the requirements of the machine directive, upon verification higher loads are possible. Several heights of optional spacers are also prepared, as well as appropriate moving tables in order to accurately position the sample environment. The base plate for the system will be aluminum EUR-pallets with an adapted three-point mounting integrated into the pallets.

In addition to those verifications, the shielding of the sample cave has been designed to avoid magnetic disturbances of a ^{3}He-analyzer for polarized neutrons. In order to achieve this, the magnetic permittivity of all walls is limited to $\mu_r \leq 1.2$, on average, and never exceeds $\mu_r = 1.8$.

Additionally, the 3 × 3 m^2 large sample area will feature a false floor, below which cabling and hoses for experimental connections can be routed. The defining sample slit of the collimation will be mounted on an extendable nose in order to minimize the free flight path of the neutrons in air.

Figure 8 details the setup. Please note that the routing of hoses and cables from top as well as from the side is also foreseen, in order to route supply media and controls to the sample environment with a closed shielding. Access from the top is intended to facilitate working with cold fingers in cryostats or other, similar top-heavy, sample equipment. For a safe handling of those sample environments also, a crane is foreseen.

The top of the generic sample environment is a breadboard of the same size as the EUR-pallet. There, all equipment can be mounted on the threaded grid.

The achievable flux at sample position is 7.7×10^8 neutrons s^{-1}cm^{-2} with 4 m collimation length. This is based on the McStas simulation of the instrument, as described in this manuscript.

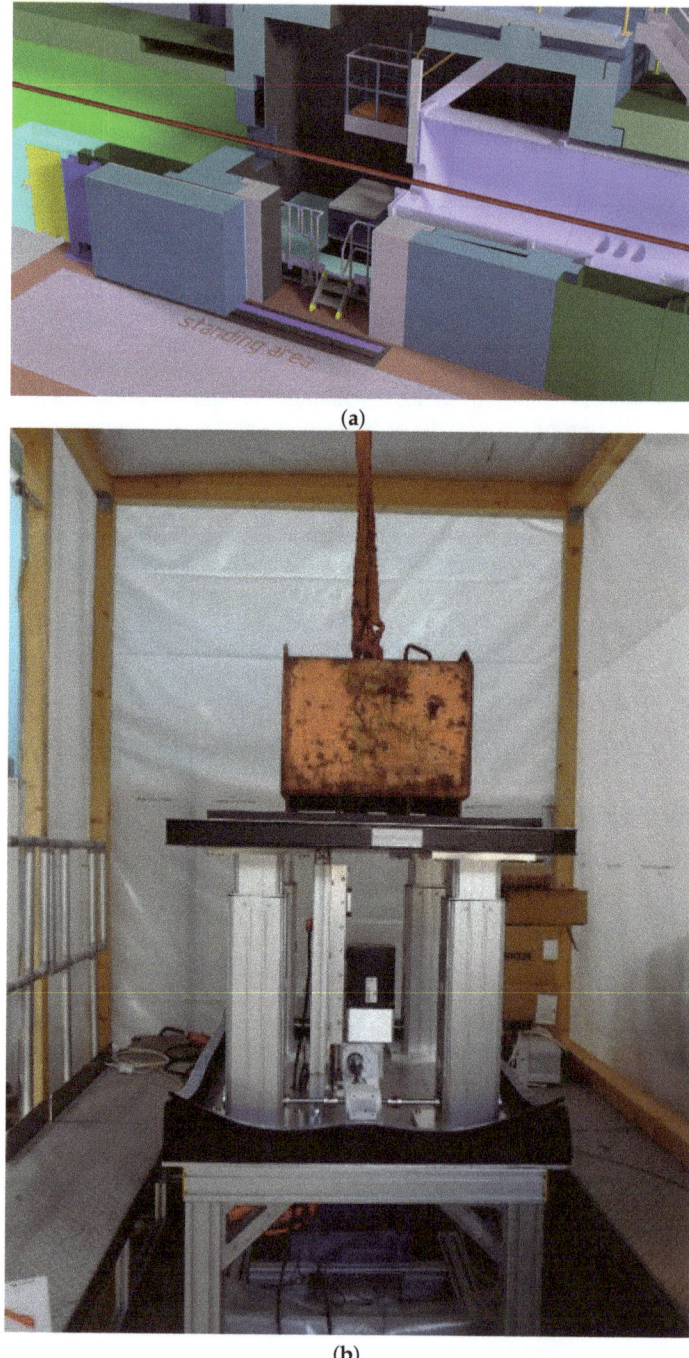

Figure 8. (a) CAD cut of the sample area, neutron beam is indicated in red. (b) Photograph of the sample stage and false floor in a mockup of the sample area, as seen from the detector tube. Test load on table is 732 kg, not affecting the positioning accuracy of 0.1 mm.

2.7. Detector

The detector used for SKADI will be a SoNDe-type pixelated scintillation detector [7,18–20]. Apart from the high-flux capability necessary [21], especially in the area of the primary beam, this detector also provides an excellent scalability and geometric freedom arranges the available active surface area as needed. In our previous publications, we also tested the sensitivity of our detector system, which is on the order of 90% for neutrons with a wavelength of $\lambda = 3$ Å.

In order to achieve a high simultaneous Q-range, SKADI will be equipped with three SoNDe type detectors, two of which are mounted on a movable carriage and have an aperture for low-angle scattering. Figure 9 shows the proposed arrangement of detector area. Here, it should be noted that the center hole is always 20×20 cm^2 and the distance ratio between the forward and middle detector, as measured from the sample, is always 1/5. The central detector is always at the same distance as the collimation setting to ensure an ideal mapping of the sample aperture on the detector, while the rear detector is fixed in the back, outside of the detector tank at 20 m. While, for some settings, there may be geometrical gaps, there is never a gap in Q-space due to the wavelength band used for the scattering experiment. In addition, the rear detector also serves as a beam monitor. A semitransparent beamstop will allow for investigating the full scattering down to the lowest angle, as the primary beam is only attenuated and never blocked.

The pixel size throughout the detector is 6×6 mm^2 everywhere, apart for the 20 m position, where the resolution is 3×3 mm^2, in order to allow for an optimal resolution. There are several considerations going affecting the choice of pixel size. In fundamental terms, the smaller pixels allow for a higher count rate, since the used detector technology has a roughly constant dead time per pixel. Thus, decreasing pixel size and increasing pixel number linearly increases the achievable count rate. Furthermore, on most current day instruments, slit sizes of roughly 1 cm in both directions are chosen as a sample blind, while the final collimation blind before that is on the order of a few centimeters in both directions. Using those numbers, the SKADI detector would not increase the achievable resolution, since the beam size gives the limit. However, for SKADI with the higher flux, smaller apertures will probably be more commonly used. Additionally, when considering the VSANS data that were recorded with the VSANS prototype that will be used by SKADI (Figure 6), it is already now apparent that the detector needs a higher resolution, in order to take full advantage of the better flux and optics.

The front and middle detector will be in a box at atmospheric pressure and located inside the vacuum tank, which is 20 m long and 2.2 m in diameter and made from aluminum [22]. Vacuum hoses in two cable chains will be required to bring the air from outside the vacuum tank to those air boxes. The cooling of the electronics and the signal out will also be routed through those hoses. Because of the 1/5 ratio of their relative distances from the sample, both of the detectors will have their own motorized support and share a common guiding rail.

(a)

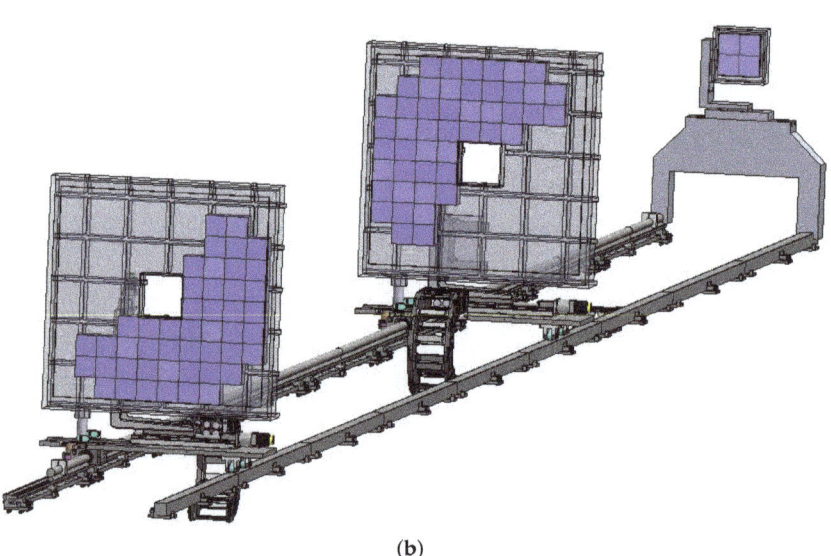

(b)

Figure 9. (**a**) Photograph of a 2 × 2 module of a SoNDe type detector. Each tile is 10 × 10 cm in size. (**b**) Arrangement of active area tiles, as shown in (**a**) for the complete detector setup.

3. Summary and Scientific Performance

The overall goal of SKADI is to cater to a wide range of scientific needs in order to allow for most conceivable small-angle neutron scattering experiments. This is achieved by a versatile sample area, a well adaptable collimation, as well as a high-flux capable detector for single shot experiments of kinetic transitions in samples. This combination should allow excellent experiments, resulting in high-profile scientific investigations and publications.

Those capabilities are accompanied by a world leading high flux at sample position and an excellent resolution down to very low Q-values around several times 10^{-4} Å^{-1} in a pin-hole setting and even lower using the VSANS setup. The upper Q limit is well above 1 Å^{-1}, depending on the specific setting. In between those limits, dynamic Q-range of $Q_{max}/Q_{min} \approx 1000$ are feasible, which makes single shot experiments over a wide range of length scales in the sample feasible.

This could be achieved by thoroughly designing SKADI from the neutron extraction to the detector with scientific performance and applicability in mind. This means that each component was first simulated as a standalone setup with a generic neutron flux, until the desired performance criteria where met. Afterwards, the specific component was simulated in tandem with the rest of the instrument and the model of the ESS moderator to assess a realistic scenario. While doing this, the pertinent parameters, such as the m-values of the guides, as well as the length and position of components, were varied in order to make sure the configuration was in a stable performance maximum. The SoNDe detector was finally developed specifically for use in a SANS instrument [7,18–20].

Summarizing this procedure, for SKADI there are a few key properties of such a SANS instrument at a high brilliance spallation source, such as the ESS.

Starting at the neutron extraction, with the necessity to displace the primary beam in order to avoid a direct view of the moderator from the sample position, the general approach of using a long curved guide would have resulted in either a very long bunker or an extremely poor transmission. This can be mitigated using a channeled bender, but the technical limitations on lifetime in a hard radiation environment and reflectivity strongly limit that approach. Additionally, mostly short wavelength and high divergencies profit from such a setup, which are not the most profitable neutrons to transport for a SANS experiment. Using the deflector setup, a twice reflected direct image of the moderator can be transported. By choosing the angles and coating appropriately, even a low wavelength cutoff can be obtained, which is both beneficial for the experiment itself, as well as shielding of the rest of the instrument.

Continuing to the chopper setup, it is clear that a time of flight instrument needs at least two choppers to define the wavelength band and prevent the frame overlap from secondary neutrons. The acceptable wavelength band is defined by the farthest detector position, and an optimal position for the first chopper is as far upstream as technically feasible, since this removes background farther from the detector. In combination, those two requirements then also define the position of the second chopper. The high resolution choppers, which still can be installed on SKADI as an upgrade, have advantages, as shown in this manuscript, however are not strictly necessary for a high performance SANS instrument and would be most profitable in the case of weak signals, a problematic signal to noise ratio, or a combination thereof.

Regarding the polarization setup, SKADI features the optimal solution both for long and short wavelengths, as well as the possibility to later upgrade for either even longer wavelength or any other optical component that might be considered to be useful. Together with the spin flipper and the ^{3}He-analyzer that can be installed at the sample position, this gives SKADI the capability to both investigate magnetic samples, or, more generally, spin dependent scattering, for all orientations, as well as the possibility to use polarization for the suppression of spin-incoherent background, which might prove useful for weakly scattering samples, where simply prolonging the measurement does not necessarily improve the data anymore.

In terms of sample environment, there is already now a lively community working of developing their own custom sample environments for SKADI. This takes advantage of the open sample environment system, which allows simple access to the sample cave with predefined kinetic mounts and other interfaces. This will both facilitate the design of custom sample environments by other users, as well as shorten the setup times between beamtimes, even for complex sample environments, since they can be prepared to a very high degree off the instrument, before the actual experiment starts.

Finally, the SoNDe detector is a completely new development, which is aimed to allow the use of the high brilliance of the ESS to capacity. The semitransparent beamstop will allow SKADI to measure down to virtually $Q = 0$ Å^{-1}, only being limited by the pixel size. In addition, no attenuation of the beam is necessary, since the SoNDe detector will be linear, even for the highest intensities available. Furthermore, the free positioning of the detectors will allow an optimal selection of resolution, collimation, and flux for any experimental setup.

Taken together, all of these features present SKADI as one of the world leading SANS instruments and an excellent addition to the ESS instrument suite.

Author Contributions: Conceptualization S.J., S.D., R.H., R.E. and H.F. (Henrich Frielinghaus); Methodology S.J.; Technical designs S.J., H.F. (Henrich Frielinghaus), T.K., R.H., S.D., P.L., A.G., S.B., R.E. and H.F. (Herbert Feilbach); writing—original draft preparation S.J.; writing—review and editing S.J., A.C., S.D., T.K., H.F. (Herbert Feilbach), P.L., R.H., A.G., S.B., R.H., H.F. (Henrich Frielinghaus), S.F. and P.M.-B.; supervision S.F. and P.M.-B.; project supervision S.J.; funding acquisition S.J. and H.F. (Henrich Frielinghaus). All authors have read and agreed to the published version of the manuscript.

Funding: R.K., T.H., P.M.-B. acknowledges funding by the BMBF project "Flexiprobe" (grant number 05K16WOA).

Conflicts of Interest: The authors declare no conflict of interest.

References

1. Peggs, S.; Kreier, R.; Carlile, C.; Miyamoto, R.; Påhlsson, A.; Trojer, M.; Weisend, J.G. (Eds). *ESS Technical Design Report*; European Spallation Source: Lund, Sweden, 2013.
2. Andersen, K.H.; Argyriou, D.N.; Jackson, A.J.; Houston, J.; Henry, P.F.; Deen, P.P.; Toft-Petersen, R.; Beran, P.; Strobl, M.; Arnold, T.; et al. The instrument suite of the European Spallation Source. *Nuc. Instrum. Meth. Phys. Res. Sec. A* **2020**, *957*, 163402. [CrossRef]
3. Jaksch, S.; Martin-Rodriguez, D.; Ostermann, A.; Jestin, J.; Pinto, S.D.; Bouwman, W.; Uher, J.; Engels, R.; Frielinghaus, H. Concept for a time-of-flight Small Angle Neutron Scattering instrument at the European Spallation Source. *Nuc. Instrum. Meth. Phys. Res. Sec. A* **2014**, *762*, 22–30. [CrossRef]
4. Widmann, T.; Kreuzer, L.P.; Kühnhammer, M.; Schmid, A.J.; Jaksch, S.; Schneider, H.; Hiess, A.; von Klitzing, R.; Hellweg, T.; Müller-Buschbaum, P. Flexible sample environment for the investigation of soft matter thin films for implementation at the European Spallation Source: The GISANS setup. *Appl. Sci.* **2021**, *11*, in press.
5. Kühnhammer, M.; Widmann, T.; Kreuzer, L.; Schmid, A.; Wiehemeier, L.; Frielinghaus, H.; Jaksch, S.; Schneider, H.; Hiess, A.; Müller-Buschbaum, P.; et al. Flexible sample environments for the investigation of soft matter at the European Spallation Source: Part III – The macroscopic foam cell. *Appl. Sci.* **2021**, *11*, in press.
6. Schmid, A.; Wiehemeier, L.; Jaksch, S.; Schneider, H.; Hiess, A.; Bögershausen, T.; Widmann, T.; Kreuzer, L.; Kühnhammer, M.; Frielinghaus, H.; et al. Flexible sample environments for the investigation of soft matter at the European Spallation Source: Part I—The in situ SANS/DLS setup. *Appl. Sci.* **2021**, *11*, in press.
7. Rofors, E.; Pallon, J.; Al Jebali, R.; Annand, J.; Boyd, L.; Christensen, M.; Clemens, U.; Desert, S.; Elfman, M.; Engels, R.; et al. Response of a Li-glass/multi-anode photomultiplier detector to focused proton and deuteron beams. *Nuc. Instrum. Meth. Phys. Res. Sec. A* **2020**, *984*, 164604. [CrossRef]
8. Brandl, G. 2018. NICOS 2 Available online: https://www.openhub.net/p/nicos2 (accessed on 16 April 2021).
9. Willendrup, P.K.; Lefmann, K. McStas (i): Introduction, use, and basic principles for ray-tracing simulations. *J. Neut. Res.* **2020**, *22*, 1–16. [CrossRef]
10. Schanzer, C.; Böni, P.; Schneider, M. High performance supermirrors on metallic substrates. *J. Phys. Conf. Ser.* **2010**, *251*, 012082. [CrossRef]
11. Jaksch, S. Considerations about chopper configuration at a time-of-flight SANS instrument at a spallation source. *Nuc. Instrum. Meth. Phys. Res. Sec. A* **2016**, *835*, 61–65. [CrossRef]
12. Feoktystov, A.V.; Frielinghaus, H.; Di, Z.; Jaksch, S.; Pipich, V.; Appavou, M.S.; Babcock, E.; Hanslik, R.; Engels, R.; Kemmerling, G.; et al. KWS-1 high-resolution small-angle neutron scattering instrument at JCNS: current state. *J. Appl. Crystallogr.* **2015**, *48*, 61–70. [CrossRef]
13. Abbas, S.; Désert, S.; Brûlet, A.; Thevenot, V.; Permingeat, P.; Lavie, P.; Jestin, J. On the design and experimental realization of a multislit-based very small angle neutron scattering instrument at the European Spallation Source. *J. Appl. Crystallogr.* **2015**, *48*, 1242–1253. [CrossRef]

14. Desert, S.; Thevenot, V.; Oberdisse, J.; Brulet, A. The new very-small-angle neutron scattering spectrometer at Laboratoire Léon Brillouin. *J. Appl. Crystallogr.* **2007**, *40*, s471–s473. [CrossRef]
15. Brûlet, A.; Thévenot, V.; Lairez, D.; Lecommandoux, S.; Agut, W.; Armes, S.P.; Du, J.; Désert, S. Toward a new lower limit for the minimum scattering vector on the very small angle neutron scattering spectrometer at Laboratoire Leon Brillouin. *J. Appl. Crystallogr.* **2008**, *41*, 161–166. [CrossRef]
16. Guinier, A.; Fournet, G. *Small-Angle Scattering of X-rays*; Wiley: New York, NY, USA; Chapman and Hall: London, UK, 1955.
17. Roe, R.J. *Methods of X-ray and Neutron Scattering in Polymer Science*; Oxford University Press on Demand: Oxford, UK, 2000.
18. Jaksch, S.; Engels, R.; Kemmerling, G.; Gheorghe, C.; Pahlsson, P.; Désert, S.; Ott, F. Cumulative Reports of the SoNDe Project July 2017. *arXiv* **2017**, arXiv:1707.08679.
19. Jaksch, S.; Engels, R.; Kemmerling, G.; Clemens, U.; Désert, S.; Perrey, H.; Gheorghe, C.; Fredriksen, A.; Øya, P.; Frielinghaus, H.; et al. Recent Developments SoNDe High-Flux Detector Project. In Proceedings of the International Conference on Neutron Optics (NOP2017), Nara, Japan, 5–8 July 2017.
20. Rofors, E.; Mauritzson, N.; Perrey, H.; Jebali, R.A.; Annand, J.; Boyd, L.; Christensen, M.; Clemens, U.; Desert, S.; Engels, R.; et al. Response of a Li-glass/multi-anode photomultiplier detector to collimated thermal-neutron beams. *arXiv* **2020**, arXiv:2010.06347.
21. Kanaki, K.; Klausz, M.; Kittelmann, T.; Albani, G.; Cippo, E.P.; Jackson, A.; Jaksch, S.; Nielsen, T.; Zagyvai, P.; Hall-Wilton, R. Detector rates for the Small Angle Neutron Scattering instruments at the European Spallation Source. *arXiv* **2018**, arXiv:1805.12334
22. Chaboussant, G.; Désert, S.; Lavie, P.; Brûlet, A. PA20: A new SANS and GISANS project for soft matter, materials and magnetism. *J. Phys. Conf. Ser.* **2012**, *340*, 012002. [CrossRef]

Article

Flexible Sample Environments for the Investigation of Soft Matter at the European Spallation Source: Part I—The In Situ SANS/DLS Setup

Andreas Josef Schmid [1], Lars Wiehemeier [1], Sebastian Jaksch [2], Harald Schneider [3], Arno Hiess [3], Torsten Bögershausen [3], Tobias Widmann [4], Julija Reitenbach [4], Lucas P. Kreuzer [4], Matthias Kühnhammer [5], Oliver Löhmann [3,5], Georg Brandl [2], Henrich Frielinghaus [2], Peter Müller-Buschbaum [4,6], Regine von Klitzing [5] and Thomas Hellweg [1,*]

1 Physikalische und Biophysikalische Chemie, Fakultät f. Chemie, Universität Bielefeld, Universitätsstr. 25, 33615 Bielefeld, Germany; andreas.josef.schmid@rwth-aachen.de (A.J.S.); lars.wiehemeier@uni-bielefeld.de (L.W.)
2 Jülich Centre for Neutron Science JCNS at Heinz Maier-Leibnitz Zentrum (MLZ), Forschungszentrum Jülich GmbH, Lichtenbergstr. 1, 85748 Garching, Germany; s.jaksch@fz-juelich.de (S.J.); g.brandl@fz-juelich.de (G.B.); h.frielinghaus@fz-juelich.de (H.F.)
3 European Spallation Source ERIC, P.O. Box 176, 221 00 Lund, Sweden; harald.schneider@ess.eu (H.S.); arno.hiess@ess.eu (A.H.); torsten.bogershausen@esss.se (T.B.); loehmann@fkp.tu-darmstadt.de (O.L.)
4 Lehrstuhl f. funktionelle Materialien, Physik Department, Technische Universität München, James-Franck-Str. 1, 85748 Garching, Germany; Tobias.widmann@ph.tum.de (T.W.); julija.reitenbach@ph.tum.de (J.R.); lucas.kreuzer@ph.tum.de (L.P.K.); muellerb@ph.tum.de (P.M.-B.)
5 Institut für Physik Kondensierter Materie, Technische Universität Darmstadt, Hochschulstr. 8, 64289 Darmstadt, Germany; kuehnhammer@fkp.tu-darmstadt.de (M.K.); klitzing@smi.tu-darmstadt.de (R.v.K.)
6 Heinz Maier-Leibnitz Zentrum (MLZ), Technische Universität München, Lichtenbergstr. 1, 85748 Garching, Germany
* Correspondence: thomas.hellweg@uni-bielefeld.de

Abstract: As part of the development of the new European Spallation Source (ESS) in Lund (Sweden), which will provide the most brilliant neutron beams worldwide, it is necessary to provide different sample environments with which the potential of the new source can be exploited as soon as possible from the start of operation. The overarching goal of the project is to reduce the downtimes of the instruments related to changing the sample environment by developing plug and play sample environments for different soft matter samples using the same general carrier platform and also providing full software integration and control by just using unified connectors. In the present article, as a part of this endeavor, the sample environment for in situ SANS and dynamic light scattering measurements is introduced.

Keywords: dynamic light scattering; small angle neutron scattering; instrumentation; microgels

1. Introduction

The new European Spallation Source [1] (ESS) in Lund (Sweden) will provide the most brilliant neutron beams worldwide. To make efficient use of this source, it is necessary to provide different sample environments with which the potential of the ESS can be exploited as soon as possible from the start of operation. Accordingly, in our opinion, now is the right time to construct and build the sample environments with the construction of "Day 1" instruments [2]. For this purpose, the expertise of three university-based groups from the field of polymer and colloid research was bundled with construction teams of the SKADI [3] and the LOKI [4] instrument and also with the sample environment group of the ESS. Three different "Day 1" sample environments were developed that use a common universal carrier system. This system was intended to enable maximum flexibility with minimal downtime caused by changing the sample environment. In addition, it should be possible to partially automate the conversion, at least in perspective, since higher radiation is to be expected at the sample location. The following three sample environments were developed as part of the joint project:

- Environment for in situ dynamic light scattering (DLS) combined with small angle neutron scattering;
- Environment for small-angle neutron scattering under grazing incidence (GISANS) [5];
- Environment for neutron scattering on foams [6].

In the first expansion step, these sample environments are to be used on ESS instruments that are optimized for elastic small angle neutron scattering (SANS) for volume samples and GISANS and reflectometry on interface samples. The necessary tests during the development phase were carried out on instruments from the MLZ (Heinz Maier–Leibnitz Zentrum) such as KWS-1 [7] and REFSANS [8]. An innovative concept for a modular sample geometry for the ESS was developed and technically implemented. In particular, the specific demands at the ESS are included in the planning.

In the project, the common model systems used to benchmark all sample environments are so-called smart microgels which are particles with tunable interaction potential. They are called "smart" due to their ability to respond to external stimuli leading to a steadily increasing number of studies dealing with these fascinating materials [9]. The major scientific question of this collaborative research endeavor addresses the interactions of these soft deformable particles in the volume phase and at interfaces. This is extremely important both with regard to the fundamental understanding of the interactions between the particles [10,11] and for the transfer to the life sciences, as well as for technical applications in new technologies [12–15].

The present work focuses on the sample environment for in situ dynamic light scattering (DLS) measurements in combination with small angle neutron scattering. SANS is a valuable method for the investigation of soft matter. Due to the high sensitivity of SANS for organic samples compared to X-ray scattering and the possibility to selectively set the contrast of the molecular species by selecting the isotopic composition of samples and solvent (so-called contrast matching), it is an extremely valuable method in many soft matter fields like studies of self-assembly [16,17], polymer melts [18,19], colloidal polymers such as microgels [20–27], micelles [28,29] and bio-macromolecules like proteins [30,31] or vesicles from lipids [17,32]. Due to the relatively low intensity of neutron beams, often rather long measurements, e.g. compared to light scattering are required to gain data of high quality.

Due to the combination of long measurements and irradiation with neutrons, a stable sample is not always guaranteed. Biological samples might be especially prone to degradation. Therefore, monitoring the sample stability during extended measurements with in situ DLS has recently gained popularity [33–37]. In this work, the components of the sample environment designed for the simultaneous measurement of SANS and DLS at the ESS with a focus on portability and the accommodation of many samples are presented (Section 2). A detailed description of the complete setup, the control software layout and the DLS data analysis can be found in the Section 3.

2. Sample Environment Components

In light of the high-intensity neutron source ESS, our setup features a sample changer for up to 39 samples with three individually temperature controlled compartments.

2.1. Sample Rack/Sample Magazines

The sample rack (Figure 1) holds three magazines for up to 13 standard 1 mm cuvettes (external dimensions: 52 mm · 12.5 mm· 3.5 mm) each. The cuvettes can be flushed with dried air to prevent condensation when measuring at low temperatures. Each of the magazines has a separate connector (CBI, Stäubli, Pfäffikon, Switzerland) for temperature control fluid and is insulated with PEEK, so that three individual temperatures can be used simultaneously, provided enough thermostats are present. Within each magazine, three positions can be equipped with in-cell Pt-100 thermometers (P0K1.161.2K.Y.5000-4.S, Innovative Sensor Technology IST AG, Ebnat-Kappel, Schweiz). The remaining cuvette temperatures can then be interpolated. The Pt-100 thermometers are read out with a temperature monitor (Model 224, Lake Shore Cryotronics, Westerville, OH, USA). The sample holder temperature is controlled with fluid pumped by a thermostat (FP50-HL, Julabo, Seelbach, Germany).

Figure 1. Photograph of the sample changer. The sample changer consists of a rack for 3 magazines each holding up to 13 cuvettes. Each magazine's temperature can be controlled individually. The open sides can be flushed with dried air to prevent condensation, when measuring at low temperatures. This part of the setup can also be used for standard SANS experiments without DLS. The red arrows indicate the cuvette positions, that were used for the temperature measurements in Figure 2.

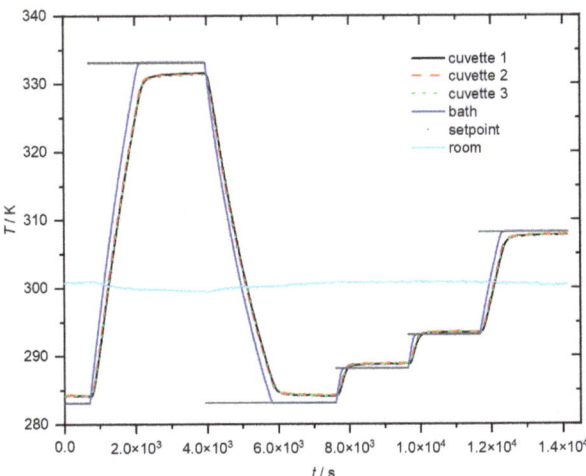

Figure 2. Comparison of set-point temperature, bath temperature and the temperature in three different cuvettes inside the sample holder plotted against time.

The sample aperture is 9 mm · 20 mm for the incident beam and 10 mm · 25 mm on the back. The largest scattering angle possible with this sample environment is 42.5° in the vertical direction. However, it should be noted that the possible scattering angle also depends on the actual configuration of the SANS instruments, e.g., the neutron spot size. However, the sample-to-aperture distance will be a compromise to allow for the incident light to reach the sample cells. The sample environment can be moved along the Y- and Z-axis (in the instrument coordinate system) to change the sample in the neutron beam and laser focus. This is accomplished with two linear translation stages (LS-180, Physik Instrumente, Karlsruhe, Germany).

Temperature Control

Precise and reproducible temperature control is an important issue when working with soft matter samples, due to the influence on molecular interactions and the sample structure. Therefore, the sample holder constructed for this instrument is characterized regarding the temperature control capabilities. The sample environment's temperature can be controlled with various fluids, allowing a temperature range from 253.15 K to 393.15 K, depending on the chosen fluid.

An example of the temperature control characteristics is shown in Figure 2. Here, the set-point temperature, the bath temperature and the temperatures in three different cuvettes filled with water in one magazine (see Figure 1) are plotted against the time. It can be seen that the cuvette temperatures follow the bath temperature rapidly. The cuvette temperatures reach 1 K of their equilibrium temperatures after 1 min and 0.1 K of the equilibrium temperature within 3 min after the bath reached the setpoint temperature even for large temperature steps (50 K). The thermostat takes ca. 30 min to reach the setpoint for cooling 50 K from 333.1 K to 283.1 K. Heating from 283.1 K to 333.1 K takes 23 min. However, smaller temperature steps, from 283.1 K to 288.1 K only take 3 min. These times are dependent on the thermostat used and are only valid for the thermostat (FP50-HL, Julabo, Seelbach, Germany) applied here.

The sample temperature can be measured using a PT-100 thermometer inside three cuvettes for each magazine. Figure 2 shows that all three cuvette temperatures are nearly identical. This indicates a homogeneous temperature distribution for all samples in one magazine. However, to correct even for these slight deviations, the sample temperature for all other cuvettes can be interpolated from the three measured temperatures in each magazine.

2.2. Light Scattering

Dynamic light scattering is a well established method for the characterization of particles in suspension by probing their diffusive behavior [38].

For the dynamic light scattering setup inspired by [36,37], the light scattering plane is inclined by 20° relative to the SANS plane, to prevent problems arising from reflections on the cuvettes. A scheme of the optical setup can be found in Figure 3.

The laser light (633 nm, 21 mW, HNL210L-EC, Thorlabs, Newton, NJ, USA) passes a shutter (SH05/M, Thorlabs, Newton, NJ, USA) and a filter wheel (FW212C, Thorlabs, Newton, NJ, USA) equipped with neutral density filters with optical densities of 0.1, 0.2, 0.3, 0.4, 0.5, 0.6, 1.0, 1.3, 2.0 and 3.0 and is coupled into a polarization-maintaining single mode fiber (P3-630PM-FC-1, Thorlabs, Newton, NJ, USA) with a fiber port (PAF-X-5-A, Thorlabs, Newton, NJ, USA). The laser light is emitted from a fiber collimator (60FC-4-M5-33, Schäfter and Kirchhoff, Hamburg, Germany) and focused onto the sample through a Glan–Thompson polarizer (GTH10M-A, Thorlabs, Newton, NJ, USA). After passing another Glan–Thompson polarizer and a laser line filter (FL632.8-1, Thorlabs, Newton, NJ, USA) the light scattered by the sample is collected by the same collimator type and coupled into a single mode optical fiber (630HP with FC/PC and FC/APC connectors, Thorlabs, Newton, NJ, USA). For the detection in pseudo-cross configuration, a fiber optic beam splitter (ALV, Langen, Germany) is used to split the scattered light in a ratio of about 50:50.

Afterwards, the light is detected into two single photon counting modules (SPCM-AQRH-14-FC, Excelitas Technologies, Waltham, MA, USA). For correlation, a field programmable gate array board (Spartan-6 FPGA, Xilinx, San José, CA, USA) is programmed analogously to a previous implementation for fluorescence correlation spectroscopy [39]. For the alignment of the setup, at first the incident laser beam is focused on the sample position. Then, an auxiliary laser is coupled into the fiber couplers of the detection side, to enable a prealignment by eye. After that, the detection can be easily adjusted to maximum scattered intensity and the quality of the correlation function.

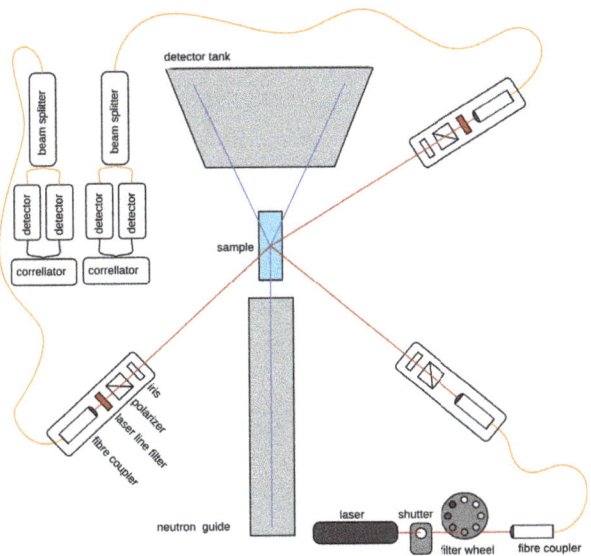

Figure 3. Scheme of the instrument components required for light scattering and the resulting beampath.

The sample environment for in situ dynamic light scattering and SANS measurements is equipped with two parallel setups for measuring dynamic light scattering at two different scattering angles θ of approximately 120° and 71°. Due to refraction on the cuvette–air and cuvette–solvent interfaces, the precise scattering angle is calculated using the refractive index n of the solvent according to Snell's law for each temperature.

To validate the correct operation of the dynamic light scattering setup, polystyrene standard spheres with a radius of 25 nm and 200 nm were measured. Figure 4 shows a autocorrelation function and the respective cumulant fit for each angle of measurement as an example. Here, sound autocorrelation functions with an intercept of about 1 indicate a well-adjusted light scattering setup. The measurement of an autocorrelation function is completed within 30 s. This enables multiple DLS experiments within the duration of a SANS experiment (some 10 min) and therefore the in situ monitoring of the sample stability.

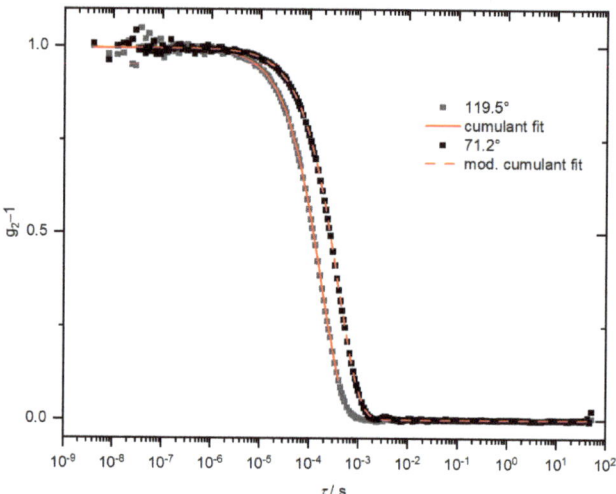

Figure 4. Two correlation functions measured simultaneously at both detection angles. The sample used here was a diluted 50 nm (diameter) polystyrene-sphere standard in H$_2$O at a temperature of 10.5 °C.

The relaxations rates Γ received from the fit can be converted into the translational diffusion coefficient D_T with the magnitude of the scattering vector $q = \frac{4\pi n}{\lambda} \sin(\frac{\Theta}{2})$. Here, λ is the wavelength, Θ is the scattering angle and n is the refractive index of the solvent.

$$D_T = \frac{\Gamma}{q^2} \qquad (1)$$

Then, the hydrodynamic radius R_H is calculated with the Boltzmann constant k_b, the solvent viscosity [40] η and the temperature T using the Stokes–Einstein equation:

$$R_H = \frac{k_b T}{6 \pi \eta D_T}. \qquad (2)$$

The hydrodynamic radii calculated this way for the particles are shown in Table 1. The measured hydrodynamic radii are consistently slightly bigger than the nominal particle size as indicated by the producer of the standard. This is a common effect in dynamic light scattering and can be attributed to the interaction of the particles with the solvent H$_2$O and maybe attractive interactions. These values are well in accordance with the measurements performed on these particles with other light scattering setups in our lab.

Table 1. Results of the dynamic light scattering measurements using two different polystyrene latex standards diluted in water with nominal radii of 25 nm and 200 nm, respectively.

	25 nm	200 nm
71°	29.9 ± 0.1 nm	209 ± 6 nm
120°	28.5 ± 0.2 nm	215 ± 3 nm

To test the dynamic light scattering setup at different temperatures, microgels are a suitable and interesting model system, due to the temperature-dependent swelling behavior. Figure 5 shows the swelling behavior of a poly N-isopropylmethacrylamide (NIPMAM) microgel crosslinked with the molecule N,N'-methylenebisacrylamide. The data were measured at both angles using the SANS/DLS sample environment. The fully swollen microgel at 12 °C exhibits a hydrodynamic radius of 150 nm. With increasing temperature, the hydrodynamic radius decreases—at first continuously, then drastically from 132 nm at 39 °C to 83 nm at 50 °C with the typical volume phase transition temperature of 45 °C for NIPMAM microgels. Within the experimental precision, the results agree perfectly with previous measurements on commercial DLS machines [26,27]. To further illustrate this agreement, Figure 5, right, shows the correlation functions of the NIPMAM microgel measured at 56 °C with the SANS/DLS setup compared to correlation functions obtained at the same q-values with a commercial setup (ALV GmbH, Langen, Germany) equipped with a Nd:YAG-laser ($\lambda = 532$ nm). The data are in excellent agreement. Moreover, no compressed exponential decays were observed for the microgel samples measured here, pointing to the absence of convection [41].

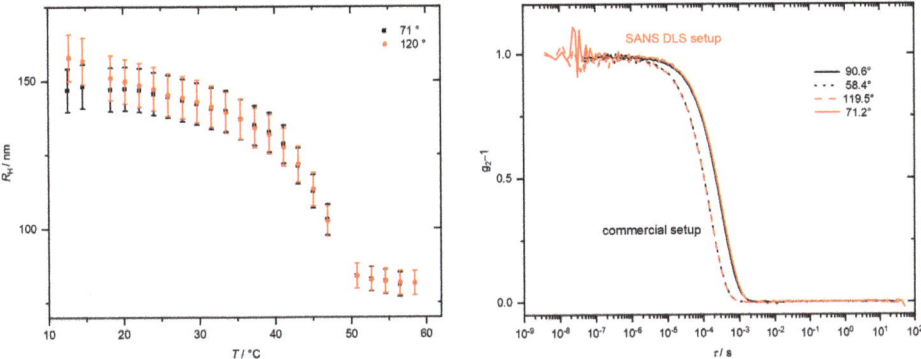

Figure 5. (**left**): Swelling behavior of a NIPMAM microgel measured with the SANS/DLS setup at both scattering angles. (**right**): comparison of autocorrelation functions of the NIPMAM microgel at 56 °C for both angles (red) compared to ACFs measured with a commercial ALV setup (black curves, measured with a different laser wavelength and therefore at different angles to achieve the same q value) at the same q-values of 0.23 nm^{-1} and 0.15 nm^{-1}.

2.3. SANS Measurements

To validate the sample environment design for SANS measurements, tests were performed at the MLZ. A sample was measured in the newly constructed sample environment and compared to neutron data previously recorded using the D11 SANS machine at the Institute Laue Langevin (ILL). Figure 6 shows the results of a measurement using the sample holder introduced in this work at the KWS-1 [7] SANS instrument at the MLZ. Figure 6, top, shows a comparison of measurements of the same NIPMAM microgel as investigated in Figure 5, performed as the reference at the ILL using the instrument D11 and with the new sample environment at MLZ. The scattering curves show a satisfactory agreement. The slight deviation at low q can be attributed to resolution effects.

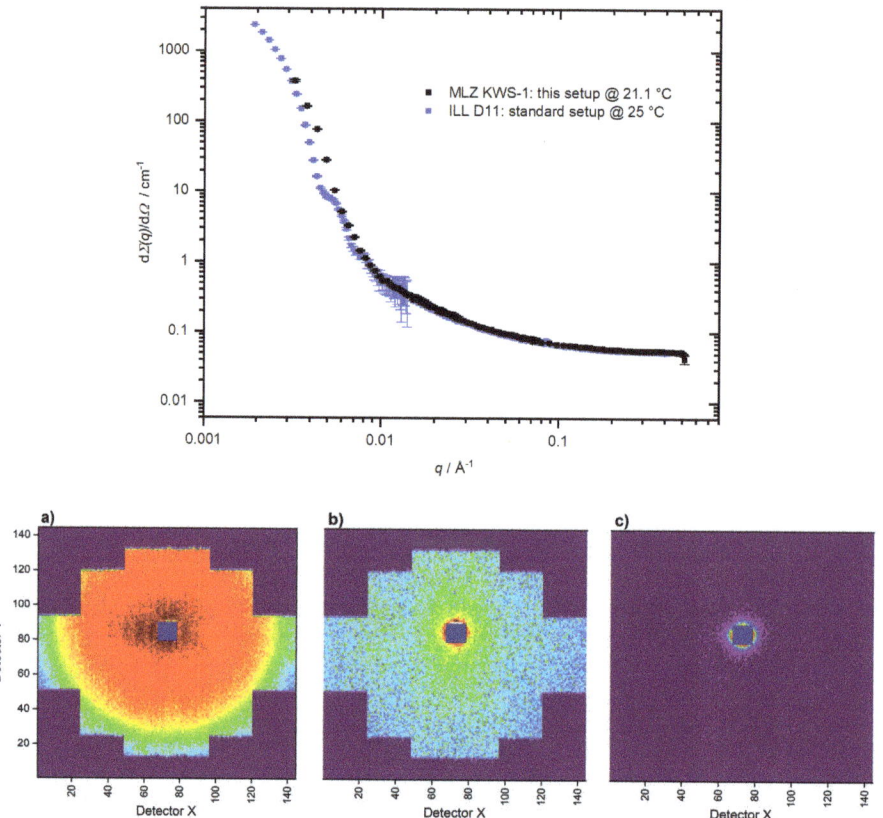

Figure 6. (**top**): SANS data measured employing the new sample environment from this work using the KWS1 machine at MLZ compared to SANS measurements made with the ILL standard setup at the D11 instrument. For both measurements, the same NIPMAM microgel sample as for the DLS measurements was used. (**bottom**): detector images for an empty cell measurement at 2 m (**a**), 5 m (**b**) and 20 m (**c**). The reflections seen here, are caused by the large distance between the nose of the collimation system and the sample cells.

Figure 6, bottom, shows detector images of empty-cell measurements at 2 m, 5 m and 20 m sample–detector distances. Especially for the detector image of the 5 m measurement, an inhomogeneous scattering can be observed. This behavior is attributed to reflections on the sample holder and the walls of the cuvettes which is caused by the rather large distance between the final window of the collimation system and the sample cells and the not fully adapted collimation. This large distance is due to spatial constraints of the sample environment which was not optimized for the KWS-1 machine.

Furthermore, a rather strong background scattering can be observed. This is also due to the relatively long air path which we had to take in the test measurements. This causes the increased background scattering. However, this will not be a problem for the instruments SKADI and LOKI at ESS, due to the optimization of the sample environment for these SANS instruments.

3. The Fully Assembled Sample Environment

3.1. General Remarks

The in situ SANS/DLS instrument setup was designed with portability and modularity in mind, to guarantee the compatibility with various SANS instruments at the ESS and to enable a fast change of sample environments with only minimal work necessary. For the portability of the setup, all equipment and electronics needed for the light scattering measurements were mounted in 19 inch racks inside a box. This box is constructed from aluminum profiles (Rose und Krieger, Minden, Germany) and fits into the footprint of a EUR-pallet. A photograph of the complete setup can be seen in Figure 7. As versatile interface a 120 cm · 90 cm breadboard (M-SG-34-2, Newport, CA, USA) is mounted on top of the aluminum box. The breadboard is screwed down on the aluminum box to prevent the shifting of the optical setup and to guarantee a stable transport of the complete setup e.g., by forklift or crane. For transport using a crane, a crane eye is installed on each edge of the box. This base construction is identical for all FlexiProb sample environments and can be easily moved to the instrument cave using a pallet jack.

The setup is constructed, so that it can be put into operation with only a few connections made. Power is supplied by a single 32 A IEC 60309 plug. The thermostat cycle needs to be connected and the RS232-cable needs to be connected to the thermostat. Further pressurized dried air, the laser interlock and an Ethernet connection are required for operation. The patch panel layout of the setup is suitable for connectors available at ESS instruments such as SKADI and LOKI.

Figure 7. Photograph of the complete SANS/DLS setup with carrier structure.

The general architecture of the interacting systems is described in the scheme in Figure 8. The user interacts with the setup either via the networked instrument control system (NICOS) [42,43] client or the DLS-GUI program, for direct control of the data fits. The measurement is controlled by the NICOS server, where the single NICOS devices can communicate with the input output controllers (IOCs) in the experimental physics and industrial control system (EPICS) [44] server. From the EPICS server, the hardware devices are controlled via their respective protocols, namely USB, Ethernet, an Ethernet-to-RS232-converter (EX-6034, EXSYS, Steinbach, Germany) or by the direct control of the power supply with switchable power outlets (Multibox, ANTRAX Datentechnik, Herford, Germany).

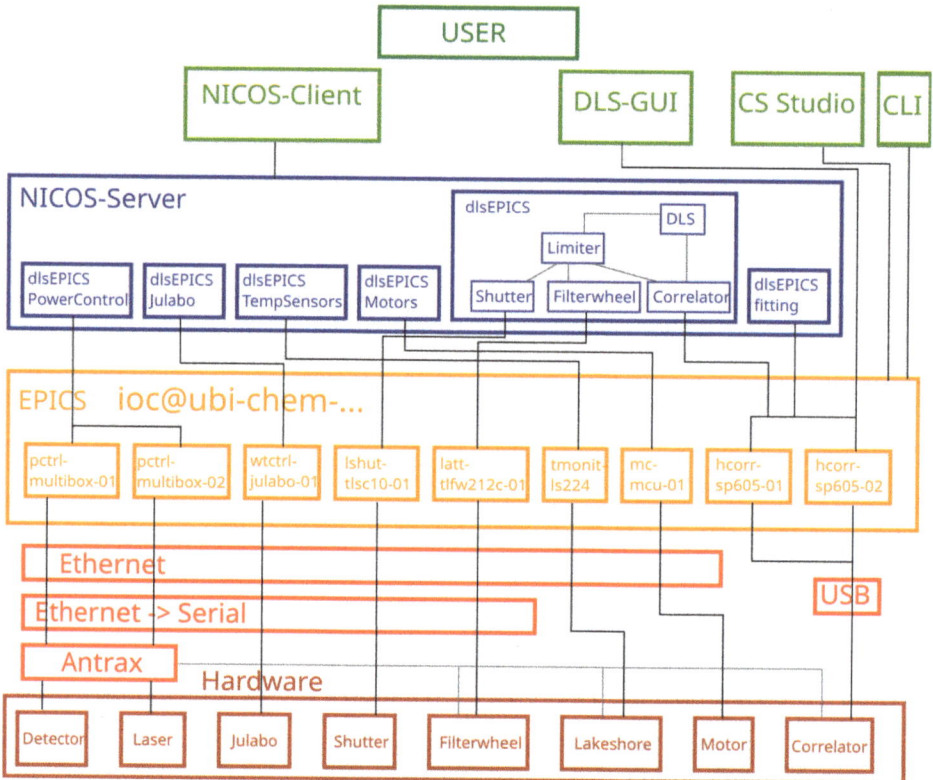

Figure 8. Schematic representation of the architecture of the setup. The user can interact via the NICOS client with the NICOS server [42,43], where the measurements are controlled. Low level-control of the devices is implemented via EPICS, that is connected via different protocols to the hardware. Alternatively, the EPICS process variables can be directly manipulated via the program CS Studio or the command line interface (CLI).

3.1.1. Safety Precautions

For laser safety, a shutter is installed in the setup. The shutter interlock will be connected to the instruments' safety environments' doors, so that no laser radiation can escape during operation. To prevent laser reflexes from the sample holder and physical contact with the cadmium, the Cd-mask in the sample holder is covered by a black polymer film. To prevent the activation of the sample environment from the neutron radiation, the sample rack and magazines are constructed from aluminum, polyether ether ketone (PEEK), and titanium screws. Boron-based neutron shielding (5 mm Mirrobor, Mirrotron, Budapest, Hungary) is applied to avert activation of the sample environment from neutron reflexes originating from the sample. For this, shielding is mounted on the side of the sample holder facing the axis of the translation unit and below the sample holder. As the optical setup could not be realized using only neutron-compatible materials, neutron shielding is also applied around the three optical towers used for light scattering.

3.1.2. Automated Data Analysis

For the analysis of the autocorrelation functions, three methods were implemented in the software. The data can be evaluated during the measurement using inverse Laplace transformation by means of the program CONTIN [45]. Moreover, the standard version of the method of cumulants [46] can be used for fits of the region of short correlation times or a modified method of cumulants can be used for larger times [47,48]. This way, the relaxation rate Γ and accordingly, the sample size and stability, can be monitored on-line during the SANS measurement.

Figure 9 shows the GUI of a custom program developed to monitor the measurements and fits during the runtime. Shown is the auto correlation function, the intensity trace (top row), the residuals from the fit and the resulting gamma distribution function (bottom row). This program bypasses the NICOS-server and receives the data directly from the correlators. Therefore, custom fits can also be tried out locally before modifying the settings for the automated fitting. Hence, the sample can be characterized in situ and deterioration can be detected during the course of the SANS measurements.

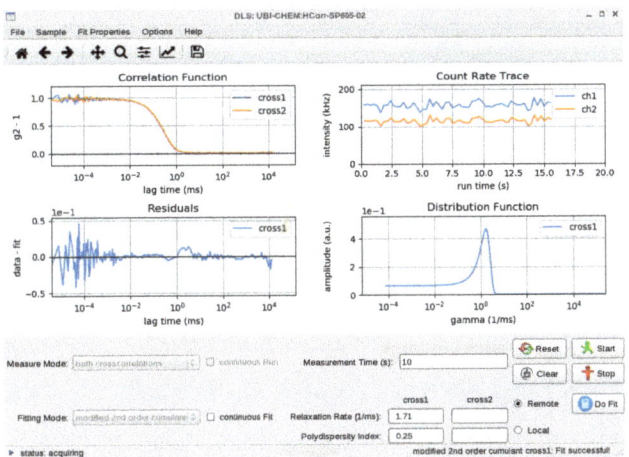

Figure 9. Application for controlling and monitoring the fits of the data during the measurements.

4. Conclusions

A new sample environment for in situ dynamic light scattering and SANS at the ESS was developed. The entire setup was transferred to MLZ and test measurements were performed. The sample environment was used to successfully measure SANS. The background problem will be solved at the ESS using a smaller distance between the individual sample cells and the end window of the collimation system. The simultaneous operation of the DLS system was proven for both configured scattering angles. The sample environments' capabilities regarding temperature control and measurement were also characterized and fulfill all specifications. The integration of all devices in both EPICS and NICOS was successfully finished, making the sample environment ready for integration and first tests at the ESS.

Author Contributions: Conceptualization, T.H., P.M.-B. and R.v.K.; SANS measurements, A.J.S. and H.F.; software, A.J.S., G.B., T.B. and L.W.; construction of the light scattering setup, A.J.S. and L.W.; construction of the carrier system: A.J.S., S.J., J.R., H.S., A.H., T.B., T.W., L.P.K., M.K., O.L. and L.W.; writing—original draft preparation, L.W. and T.H.; writing—review and editing, A.J.S., S.J., H.S., A.H., T.B., T.H., T.W., J.R., L.P.K., M.K., H.F., O.L., P.M.-B., R.v.K.; visualization, L.W.; supervision, T.H.; project administration, T.H.; funding acquisition, T.H., P.M.-B. and R.v.K. All authors have read and agreed to the published version of the manuscript.

Funding: This research was funded by the German Federal Ministry for Education and Research (BMBF) within the project "FlexiProb" sample environment grant number 05K2016.

Institutional Review Board Statement: Not applicable.

Informed Consent Statement: Not applicable.

Data Availability Statement: The data shown is available on request from the corresponding author.

Acknowledgments: We gratefully acknowledge the ESS Motion Control Group for help implementing the control of the translation stage. We gratefully acknowledge Tobias Schrader for varied fruitful discussions, Yvonne Hannapel and Simon Staringer for help during the SANS test experiments. Furthermore, we thank Claus Seidel (HHU, Düsseldorf) for the original version of the software used to operate the FPGA as a correlator. We gratefully acknowledge financial support for the publication costs by the Open Access Publication Fund of Bielefeld University.

Conflicts of Interest: The authors declare no conflict of interest.

References

1. Garoby, R.; Vergara, A.; Danared, H.; Alonso, I.; Bargallo, E.; Cheymol, B.; Darve, C.; Eshraqi, M.; Hassanzadegan, H.; Jansson, A.; et al. The European Spallation Source Design. *Phys. Scr.* **2017**, *93*, 014001. [CrossRef]
2. Andersen, K.; Argyriou, D.; Jackson, A.; Houston, J.; Henry, P.; Deen, P.; Toft-Petersen, R.; Beran, P.; Strobl, M.; Arnold, T.; et al. The instrument suite of the European Spallation Source. *Nucl. Instrum. Methods Phys. Res. Sect. A Accel. Spectrometers Detect. Assoc. Equip.* **2020**, *957*, 163402. [CrossRef]
3. Frielinghaus, P.H. *ESS Instrument Construction Proposal SKADI*; ESS: Lund, Sweden, 2017.
4. Jackson, A.J.; Kanaki, K. *ESS Construction Proposal LoKI—A Broad-Band SANS Instrument*; ESS: Lund, Sweden, 2013.
5. Widmann, T.; Kreuzer, L.P.; Kühnhammer, M.; Löhmann, O.; Schmid, A.J.; Wiehemeier, L.; Jaksch, S.; Frielinghaus, H.; Schneider, H.; Hiess, A.; et al. Flexible sample environment for the investigation of soft matter at the European Spallation Source: Part II-The GISANS setup. *Appl. Sci.* **2021**, *11*, 4036,
6. Kühnhammer, M.; Widmann, T.; Kreuzer, L.; Schmid, A.J.; Wiehemeier, L.; Frielinghaus, H.; Jaksch, S.; Bögershausen, T.; Barron, P.; Schneider, H.; et al. Flexible sample environments for the investigation of soft matter at the European Spallation Source: Part III–The macroscopic foam cell. *Appl. Sci.* **2021**, to be submitted.
7. Feoktystov, A.V.; Frielinghaus, H.; Di, Z.; Jaksch, S.; Pipich, V.; Appavou, M.S.; Babcock, E.; Hanslik, R.; Engels, R.; Kemmerling, G.; et al. KWS-1 high-resolution small-angle neutron scattering instrument at JCNS: Current state. *J. Appl. Crystallogr.* **2015**, *48*, 61–70. [CrossRef]
8. Moulin, J.F.; Haese, M. REFSANS: Reflectometer and evanescent wave small angle neutron spectrometer. *J. Large-Scale Res. Facil. JLSRF* **2015**, *1*, 9. [CrossRef]
9. Karg, M.; Pich, A.; Hellweg, T.; Hoare, T.; Lyon, L.A.; Crassous, J.J.; Suzuki, D.; Gumerov, R.A.; Schneider, S.; Potemkin, I.I.; et al. Nanogels and microgels: From model colloids to applications, recent developments, and future trends. *Langmuir* **2019**, *35*, 6231–6255. [CrossRef]
10. Scotti, A.; Denton, A.R.; Brugnoni, M.; Houston, J.E.; Schweins, R.; Potemkin, I.I.; Richtering, W. Deswelling of microgels in crowded suspensions depends on cross-link density and architecture. *Macromolecules* **2019**, *52*, 3995–4007. [CrossRef]
11. Mohanty, P.S.; Nöjd, J.; van Gruijthuijsen, K.; Crassous, J.J.; Obiols-Rabasa, M.; Schweins, R.; Stradner, A.; Schurtenberger, P. Interpenetration of polymeric microgels at ultrahigh densities. *Sci. Rep.* **2017**, *7*, 1–12. [CrossRef] [PubMed]
12. Lyon, L.A.; Meng, Z.; Singh, N.; Sorrell, C.D.; John, A.S. Thermoresponsive microgel-based materials. *Chem. Soc. Rev.* **2009**, *38*, 865. [CrossRef]
13. Menne, D.; Pitsch, F.; Wong, J.E.; Pich, A.; Wessling, M. Temperature-modulated water filtration using microgel-functionalized hollow-fiber membranes. *Angew. Chem. Int. Ed.* **2014**, *53*, 5706–5710. [CrossRef] [PubMed]
14. Uhlig, K.; Wegener, T.; He, J.; Zeiser, M.; Bookhold, J.; Dewald, I.; Godino, N.; Jaeger, M.; Hellweg, T.; Fery, A.; et al. Patterned thermoresponsive microgel coatings for noninvasive processing of adherent cells. *Biomacromolecules* **2016**, *17*, 1110–1116. [CrossRef] [PubMed]
15. Bookhold, J.; Dirksen, M.; Wiehemeier, L.; Knust, S.; Anselmetti, D.; Paneff, F.; Zhang, X.; Gölzhäuser, A.; Kottke, T.; Hellweg, T. Smart membranes by electron beam cross-linking of copolymer microgels. *Soft Matter* **2021**, *17*, 2205–2214. [CrossRef] [PubMed]
16. Bergström, M.; Pedersen, J.S.; Schurtenberger, P.; Egelhaaf, S.U. Small-angle neutron scattering (SANS) study of vesicles and lamellar sheets formed from mixtures of an anionic and a cationic surfactant. *J. Phys. Chem. B* **1999**, *103*, 9888–9897. [CrossRef]
17. Dargel, C.; Geisler, R.; Hannappel, Y.; Kemker, I.; Sewald, N.; Hellweg, T. Self-assembly of the bio-surfactant aescin in solution: A small-angle x-ray scattering and fluorescence study. *Colloids Interfaces* **2019**, *3*, 47. [CrossRef]
18. Muller, R.; Pesce, J.J.; Picot, C. Chain conformation in sheared polymer melts as revealed by SANS. *Macromolecules* **1993**, *26*, 4356–4362. [CrossRef]
19. Hammouda, B. SANS from homogeneous polymer mixtures: A unified overview. In *Advances in Polymer Science*; Springer: Berlin/Heidelberg, Germany, 1993; pp. 87–133. [CrossRef]
20. Tennenbaum, M.; Anderson, C.; Hyatt, J.S.; Do, C.; Fernandez-Nieves, A. Internal structure of ultralow-crosslinked microgels: From uniform deswelling to phase separation. *Phys. Rev. E* **2021**, *103*, 022614. [CrossRef]
21. Widmann, T.; Kreuzer, L.P.; Hohn, N.; Bießmann, L.; Wang, K.; Rinner, S.; Moulin, J.F.; Schmid, A.J.; Hannappel, Y.; Wrede, O.; et al. Hydration and Solvent Exchange Induced Swelling and Deswelling of Homogeneous Poly(N-isopropylacrylamide) Microgel Thin Films. *Langmuir* **2019**, *35*, 16341–16352. [CrossRef] [PubMed]
22. Virtanen, O.L.J.; Kather, M.; Meyer-Kirschner, J.; Melle, A.; Radulescu, A.; Viell, J.; Mitsos, A.; Pich, A.; Richtering, W. Direct monitoring of microgel formation during precipitation polymerization of N-isopropylacrylamide using in situ SANS. *ACS Omega* **2019**, *4*, 3690–3699. [CrossRef]

23. Cors, M.; Wiehemeier, L.; Hertle, Y.; Feoktystov, A.; Cousin, F.; Hellweg, T.; Oberdisse, J. Determination of internal density profiles of smart acrylamide-based microgels by small-angle neutron scattering: A multishell reverse Monte Carlo approach. *Langmuir* **2018**, *34*, 15403–15415. [CrossRef]
24. Stieger, M.; Pedersen, J.S.; Lindner, P.; Richtering, W. Are thermoresponsive microgels model systems for concentrated colloidal suspensions? A rheology and small-angle neutron scattering study. *Langmuir* **2004**, *20*, 7283–7292. [CrossRef] [PubMed]
25. Crowther, H.M.; Saunders, B.R.; Mears, S.J.; Cosgrove, T.; Vincent, B.; King, S.M.; Yu, G.E. Poly(NIPAM) microgel particle de-swelling: A light scattering and small-angle neutron scattering study. *Colloids Surfaces A Physicochem. Eng. Asp.* **1999**, *152*, 327–333. [CrossRef]
26. Wedel, B.; Zeiser, M.; Hellweg, T. Non NIPAM based smart microgels: Systematic variation of the volume phase transition temperature by copolymerization. *Z. Phys. Chem.* **2012**, *226*, 737–748. [CrossRef]
27. Wiehemeier, L.; Cors, M.; Wrede, O.; Oberdisse, J.; Hellweg, T.; Kottke, T. Swelling behaviour of core–shell microgels in H2O, analysed by temperature-dependent FTIR spectroscopy. *Phys. Chem. Chem. Phys.* **2019**, *21*, 572–580. [CrossRef]
28. Sochor, B.; Düdükcü, Ö.; Lübtow, M.M.; Schummer, B.; Jaksch, S.; Luxenhofer, R. Probing the complex loading-dependent structural changes in ultrahigh drug-loaded polymer micelles by small-angle neutron scattering. *Langmuir* **2020**, *36*, 3494–3503. [CrossRef] [PubMed]
29. Hayter, J.B. Neutron scattering from concentrated micellar solutions. *Berichte Bunsenges. Phys. Chem.* **1981**, *85*, 887–891. [CrossRef]
30. Svergun, D.I.; Richard, S.; Koch, M.H.J.; Sayers, Z.; Kuprin, S.; Zaccai, G. Protein hydration in solution: Experimental observation by x-ray and neutron scattering. *Proc. Natl. Acad. Sci. USA* **1998**, *95*, 2267–2272. [CrossRef]
31. Cousin, F.; Gummel, J.; Ung, D.; Boué, F. Polyelectrolyte-protein complexes: Structure and conformation of each specie revealed by SANS. *Langmuir* **2005**, *21*, 9675–9688. [CrossRef]
32. Sreij, R.; Dargel, C.; Geisler, P.; Hertle, Y.; Radulescu, A.; Pasini, S.; Perez, J.; Moleiro, L.H.; Hellweg, T. DMPC vesicle structure and dynamics in the presence of low amounts of the saponin aescin. *Phys. Chem. Chem. Phys.* **2018**, *20*, 9070–9083. [CrossRef]
33. Nigro, V.; Angelini, R.; King, S.; Franco, S.; Buratti, E.; Bomboi, F.; Mahmoudi, N.; Corvasce, F.; Scaccia, R.; Church, A.; et al. Apparatus for simultaneous dynamic light scattering–small angle neutron scattering investigations of dynamics and structure in soft matter. *Rev. Sci. Instrum.* **2021**, *92*, 023907. [CrossRef]
34. Nawroth, T.; Buch, P.; Buch, K.; Langguth, P.; Schweins, R. Liposome formation from bile salt–lipid micelles in the digestion and drug delivery model FaSSIFmod estimated by combined time-resolved neutron and dynamic light scattering. *Mol. Pharm.* **2011**, *8*, 2162–2172. [CrossRef] [PubMed]
35. Balacescu, L.; Vögl, F.; Staringer, S.; Ossovyi, V.; Brandl, G.; Lumma, N.; Feilbach, H.; Holderer, O.; Pasini, S.; Stadler, A.; et al. In situ dynamic light scattering complementing neutron spin-echo measurements on protein samples. *J. Surf. Investig. X-Ray Synchrotron Neutron Tech.* **2020**, *14*, S185–S189. [CrossRef]
36. Kohlbrecher, J.; Bollhalder, A.; Vavrin, R.; Meier, G. A high pressure cell for small angle neutron scattering up to 500 MPa in combination with light scattering to investigate liquid samples. *Rev. Sci. Instrum.* **2007**, *78*, 125101. [CrossRef] [PubMed]
37. Vavrin, R.; Kohlbrecher, J.; Wilk, A.; Ratajczyk, M.; Lettinga, M.P.; Buitenhuis, J.; Meier, G. Structure and phase diagram of an adhesive colloidal dispersion under high pressure: A small angle neutron scattering, diffusing wave spectroscopy, and light scattering study. *J. Chem. Phys.* **2009**, *130*, 154903. [CrossRef] [PubMed]
38. Berne, B.J.; Pecora, R. *Dynamic Light Scattering: With Applications to Chemistry, Biology, and Physics*; Dover Publications Inc.: New York, NY, USA, 2000; ISBN 0486442276.
39. Kalinin, S.; Kühnemuth, R.; Vardanyan, H.; Seidel, C.A.M. Note: A 4 ns hardware photon correlator based on a general-purpose field-programmable gate array development board implemented in a compact setup for fluorescence correlation spectroscopy. *Rev. Sci. Instrum.* **2012**, *83*, 096105. [CrossRef]
40. Cho, C.H.; Urquidi, J.; Singh, S.; Robinson, G.W. Thermal offset viscosities of liquid H2O, D2O, and T2O. *J. Phys. Chem. B* **1999**, *103*, 1991–1994. [CrossRef]
41. Gabriel, J.; Blochowicz, T.; Stühn, B. Compressed exponential decays in correlation experiments: The influence of temperature gradients and convection. *J. Chem. Phys.* **2015**, *142*, 104902. [CrossRef]
42. Brandl, G.; Felder, C.; Pedersen, B.; Faulhaber, E.; Lenz, A.; Krüger, J. NICOS–The Instrument Control solution at the MLZ. In Proceedings of the 10th International Workshop on Personal Computers and Particle Accelerator Controls, Karlsruhe, Germany, 14–17 October 2014; Number IMPULSE-2014-00016.
43. Available online: https://nicos-controls.org/ (accessed on 29 April 2021).
44. Dalesio, L.R.; Kozubal, A.J.; Kraimer, M.R. *EPICS Architecture*; Los Alamos National Lab.: Los Alamos, NM, USA, 1991.
45. Provencher, S.W. CONTIN: A general purpose constrained regularization program for inverting noisy linear algebraic and integral equations. *Comput. Phys. Commun.* **1982**, *27*, 229–242. [CrossRef]
46. Koppel, D.E. Analysis of macromolecular polydispersity in intensity correlation spectroscopy: The method of cumulants. *J. Chem. Phys.* **1972**, *57*, 4814–4820. [CrossRef]
47. Frisken, B.J. Revisiting the method of cumulants for the analysis of dynamic light-scattering data. *Appl. Opt.* **2001**, *40*, 4087. [CrossRef]
48. Hassan, P.; Kulshreshtha, S. Modification to the cumulant analysis of polydispersity in quasielastic light scattering data. *J. Colloid Interface Sci.* **2006**, *300*, 744–748. [CrossRef] [PubMed]

Article

Flexible Sample Environment for the Investigation of Soft Matter at the European Spallation Source: Part II—The GISANS Setup

Tobias Widmann [1], Lucas P. Kreuzer [1], Matthias Kühnhammer [2], Andreas J. Schmid [3], Lars Wiehemeier [3], Sebastian Jaksch [4,5], Henrich Frielinghaus [4], Oliver Löhmann [2,6], Harald Schneider [6], Arno Hiess [6], Regine von Klitzing [2], Thomas Hellweg [3] and Peter Müller-Buschbaum [1,5,*]

1 Lehrstuhl für Funktionelle Materialien, Physik Department, Technische Universität München, James-Franck-Str. 1, 85748 Garching, Germany; Tobias.widmann@ph.tum.de (T.W.); lucas.kreuzer@ph.tum.de (L.P.K.)
2 Institut für Physik Kondensierter Materie, Technische Universität Darmstadt, Hochschulstraße 8, 64289 Darmstadt, Germany; kuehnhammer@fkp.tu-darmstadt.de (M.K.); oliver.loehmann@bam.de (O.L.); klitzing@fkp.tu-darmstadt.de (R.v.K.)
3 Physikalische und Biophysikalische Chemie, Fakultät f. Chemie, Universität Bielefeld, Universitätsstr. 25, 33515 Bielefeld, Germany; andreas_josef.schmid@uni-bielefeld.de (A.J.S.); lars.wiehemeier@uni-bielefeld.de (L.W.); thomas.hellweg@uni-bielefeld.de (T.H.)
4 Jülich Centre for Neutron Science (JCNS) at Heinz Maier-Leibnitz Zentrum, Forschungszentrum Jülich GmbH, Lichtenberstraße 1, 85747 Garching, Germany; s.jaksch@fz-juelich.de (S.J.); h.frielinghaus@fz-juelich.de (H.F.)
5 Heinz Maier-Leibnitz Zentrum (MLZ), Technische Universität München, Lichtenbergstr. 1, 85748 Garching, Germany
6 European Spallation Source ERIC, P.O. Box 176, SE-221 00 Lund, Sweden; Harald.Schneider@ess.eu (H.S.); arno.hiess@ess.eu (A.H.)
* Correspondence: muellerb@ph.tum.de

Abstract: The FlexiProb project is a joint effort of three soft matter groups at the Universities of Bielefeld, Darmstadt, and Munich with scientific support from the European Spallation Source (ESS), the small-K advanced diffractometer (SKADI) beamline development group of the Jülich Centre for Neutron Science (JCNS), and the Heinz Maier-Leibnitz Zentrum (MLZ). Within this framework, a flexible and quickly interchangeable sample carrier system for small-angle neutron scattering (SANS) at the ESS was developed. In the present contribution, the development of a sample environment for the investigation of soft matter thin films with grazing-incidence small-angle neutron scattering (GISANS) is introduced. Therefore, components were assembled on an optical breadboard for the measurement of thin film samples under controlled ambient conditions, with adjustable temperature and humidity, as well as the optional in situ recording of the film thickness via spectral reflectance. Samples were placed in a 3D-printed spherical humidity metal chamber, which enabled the accurate control of experimental conditions via water-heated channels within its walls. A separately heated gas flow stream supplied an adjustable flow of dry or saturated solvent vapor. First test experiments proved the concept of the setup and respective component functionality.

Keywords: GISANS; sample environment; 3D printed; humidity chamber; thin films; SANS

1. Introduction

Neutron scattering techniques are well-established routines for the structural and molecular analysis of matter, and they cover a broad field of scientific interest [1–10]. With techniques such as neutron reflectometry (NR) [11–13], which is used for the investigation of vertical thin film compositions, or grazing-incidence small-angle neutron scattering (GISANS) [14–17], which is used for the investigation of lateral thin film morphology and its order, two powerful methods to study thin films exist. GISANS probes a large material volume by measuring the scattered signal from a neutron beam impinging under a shallow incident angle (typically below 0.5°). Therefore, good statistics over a larger sample area can be achieved while also providing additional enhancements of the surface correlated and sub-surface structural information of the thin film. Nevertheless, for thin film samples, the sampled material volume is still orders of magnitude smaller than for bulk samples. Thus, in conjunction with the available neutron flux, times in GISANS measurement are typically in the range of hours. Only a limited number of facilities worldwide provide the necessary resources and expertise for neutron scattering, and, as such, beamtime is a highly limited resource—even more so with older facilities, e.g., the BER II [17], the Orphée [18], and the JEEP II [19], being shut down.

A new facility like the European spallation source (ESS), which is under construction in Lund, Sweden, will provide an increased flux [20], which is needed for methods such as GISANS. The ESS instrument suite [21] covers a broad field of scientific interests, and, amongst others, it offers two small-angle scattering instruments suitable for investigating soft matter thin films, namely the LOKI [22] and the SKADI [23] instruments. Moreover, the ESS aims to provide a pulsed neutron beam up to 100 times more brilliant than what is currently available and would thus greatly reduce measurement times. This might allow for even more flux-dependent applications of grazing-incidence methods, e.g., in spectroscopy [24]. With an increased flux, the ratio between setup time and actual measurements increases, which makes reducing the setup times a major tool for the overall efficiency of instruments. In order to optimize setup procedures, the FlexiProb project aims for a flexible and fast sample environment exchange system. In this joint project, three independent sample environment systems addressing different scientific topics were designed by the groups of Bielefeld, Darmstadt, and Munich, with the Munich system being presented here. They share an identical base framework that enables a quick and easy exchange mechanism between the setups. It consists of a kinematic mounting system that connects to the floor, a vertical lifting table for an initial vertical alignment, and appropriate spacers with a breadboard on top. The accuracy of the kinematic mounting system is better than 0.1 mm in all directions. On top of the breadboard, the sample environment—which, in our case, consists of the various components for GISANS measurements on soft matter thin films—is assembled. All technical equipment can be installed and adjusted before the experiment and initial alignment and preparation procedures can be done off-instrument. This allows one to quickly and efficiently exchange complete and complex setups by exchanging the whole system (mounting system, lifting table, and breadboard with measurement equipment) in a single step and by reducing or even avoiding the need for adjustments and alignments after the setup is installed at the instrument.

It is crucial for the sample environment to provide stable control over temperature and humidity because various soft matter systems are highly sensitive to these parameters [25–29]. There is a wide range of humidity chambers used for controlling the vapor composition around a sample during neutron scattering experiments [30–32]. However, usually the sample size and available space around the sample is very limited, which makes the setup very specific for its intended application. In order to control the humidity in such chambers, saturated salt solutions [33–35], heated solvent reservoirs [36–38], or solvent-saturated gas flows [32,39,40] are usually used.

In this work, we developed a sample environment that included a spherical 3D-printed humidity chamber for sample placement, a gas flow control assembly for atmospheric control and vapor mixing, a set-up for in situ spectral reflectance measurements, and a goniometer assembly for motion control and accurate sample alignment. All components were assembled on a 1200 × 900 mm^2 optical breadboard and were tested for their functionality. Functional activity with neutrons has already been tested at the KWS-1 at the MLZ, Garching, Germany [41,42]. The full testing of the setup, including the off-instrument alignment and preparation procedures, are in place and will be verified as soon as the ESS instruments become operational.

2. Sample Environment Components

2.1. Humidity Chamber

We designed a spherical 3D-printed humidity metal chamber that provides precise control over temperature and, in conjunction with the gas flow assembly, humidity while offering a relatively large volume that can accommodate large samples and additional, interchangeable equipment. It is an upgraded version of a previous design of ours, with an improved control over environmental parameters and an easier general handling of the chamber [43]. The spherical design depicted in Figure 1 enables a homogeneous heat distribution for isothermal and temperature switching experiments while simultaneously preventing condensation due to the lack of sharp edges and corners, which are usually critical points for condensation [44]. The chamber size was reduced to a diameter of 146 mm, with a plane-to-plane distance of the two windows of 138 mm. It is still able to accommodate samples up to a size of 70 × 70 mm^2, while the smaller overall chamber volume of 1.0 L compared to the previous volume of 1.4 L allows for faster temperature and humidity switching kinetics. The connection between chamber and lid was changed to a flat geometry in order to facilitate the accurate positioning of the chamber lid, on which the setup for spectral reflectance measurements is mounted. The windows have a height of 30 mm and a width of 90 mm to enable NR measurements. The angle between a window edge and the furthest sample edge (of a 70 × 70 mm^2 sample placed in the center) measures to 8.2°, well above the typical scattering angles needed for both NR and GISANS measurements. This ensures that the 3D printed material is not hit by the incident nor the scattered neutron beam (which would produce unnecessary background). It was shown with the previous chamber design that the 3D printed material yields a strong scattering signal when directly hit, but it generates no detectable background otherwise [43,45,46]. As such, NR and GISANS measurements with no measurable background contribution from the chamber itself were already demonstrated. Within the 10-mm-thick walls of the chamber, a pathway of a 6-mm-diameter fluidic channel for temperature control is centered and runs around the chamber, as shown in Figure 1b. The total path length is 2.2 m, thus providing ample contact area for a thorough heat transfer.

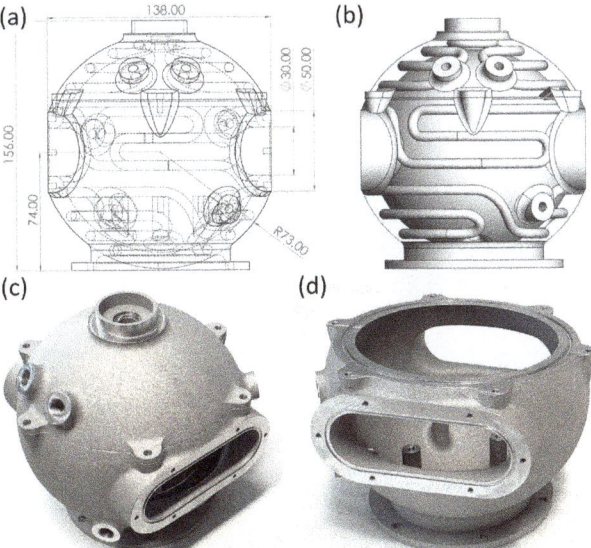

Figure 1. (**a**) Wireframe model with important measures indicated in mm. (**b**) Chamber with the outer surface hidden in order to reveal the fluid pathways embedded within the chamber walls. (**c**,**d**) Chamber printed out of AlSi$_{10}$Mg with all contact surfaces polished.

The chamber is 3D printed from an AlSi10Mg aluminum alloy powder via selective laser melting (SLM) using a metal 3D printer (EOS M290, PROTIQ GmbH, Blomberg, Germany). During the layer-by-layer printing process, the construction chamber was flooded with an inert gas to prevent the oxidation of the material. Afterwards, a heat treatment ensured the homogenization of the intermetallic phase and reduced grain boundaries within the bulk material, improving the direction-independent tensile strength of the chamber. Lastly, the contact surfaces were polished flat to ensure the airtight sealing of the chamber, thereby yielding the printed chamber and lid shown in Figure 1c,d. The lid of the chamber is designed to accommodate the optical lens system of the spectral reflectance setup, while the holes at the side of the chamber are designed as connections to either the internal liquid flow channel or the inside of the chamber, providing in and outlets for the gas flow system and additional electronics.

For measurements, the three pillars at the bottom of the chamber, visible through the chambers window in Figure 1d, are used to mount an aluminum plate with three vertical pins that can be adjusted to accommodate various sample sizes from 10×10 to 70×70 mm^2. The windows are covered by a 0.1 mm aluminum foil clamped against the windows by a 2-mm-thick aluminum counterpart. For further details on the sample holder and the various connections to the chamber, we refer to our previous publication [43].

The temperature within the chamber is controlled by a heated or cooled liquid, in the current case H_2O, flowing through the channels within the walls. It is usually adjusted between 10 and 80 °C for soft matter experiments, but it can be extended by using a different heating medium. In order to investigate the temperature distribution around the sample during experiments, we conducted temperature-switching simulations and confirmed them with corresponding laboratory experiments. For that purpose, SolidWorks flow simulation was used to simulate a heated and a cooled liquid flow through the chamber, heating it from 20 to 50 °C and cooling it from 50 to 20 °C, respectively. The liquid flow was set to start through the chamber lid before flowing via a tube to the lower part of the chamber and finally exiting at an intermediate height (see number order in the first image of Figure 2a). For the simulation, a liquid flow speed of 30 mL s^{-1}, a heat transfer coefficient of 10 W m^{-2} K^{-1}, an ambient temperature of 20 °C, and turbulence free surrounding conditions were assumed. The chosen flow speed is related to the flow speed limit of the heating liquid through the channels. Figure 2a depicts the sectional cuts through the center of the chamber obtained during the heating simulation. The cuts are shown as contour plots, coloring the heat distribution within the chamber in isothermals with a 1 °C range. We find laterally homogeneous heating layers as the temperature quickly changes over a period of around one minute.

An analogous laboratory experiment was conducted, connecting a 50 °C liquid flow to the chamber that was preheated to 20 °C. The liquid flow was provided by a refrigerated heating circulator (FP50-HL, JULABO Labortechnik GmbH, Seelbach, Germany), and the temperature was measured with a combined humidity and temperature sensor (SHT31, Sensirion AG, Steafa, Switzerland) throughout the whole experiment. Figure 2b depicts the temperature evolution at the center of the chamber over the course of the simulation and the laboratory experiment. We find a slightly slower heat evolution during the experiments, which we attribute to heat losses due to air turbulence around the chamber in the laboratory environment not being included in the simulation. Using the same settings as for the heating process, a cooling simulation and experiment were performed by switching back from the 50 °C liquid stream to a 20 °C stream. The temperature evolution at the center of the chamber is depicted in Figure 2c. Again, the evolution of the temperature extracted from the simulation and the measured temperature are in good agreement.

The smaller chamber size (compared to the previous design) and the thus potentially stronger heat discrepancies between the sample center and its edges might affect the homogeneity of the heat distribution, especially for large samples. Therefore, we measured the temperature at the chamber center (c) and 35 mm off-center (oc) in the direction of the windows (where the most heterogeneous temperature distribution was expected) over the course of the previously described temperature jump experiment. The results are depicted in Figure 3, together with the temperature at the two locations obtained from the previous simulation. The simulated temperatures show a discrepancy of less than 0.1 °C, and the curves strongly overlap. In comparison, the measured temperature closer to the chamber windows is slightly trailing behind the measured temperature at the center. However, considering the accuracy of the temperature sensors given with 0.5 °C, the differences are marginal. This supports the presence of the horizontal temperature layers that were found in the contour plots of the simulation and shows that the chamber provides a good temperature stability over the area of large samples.

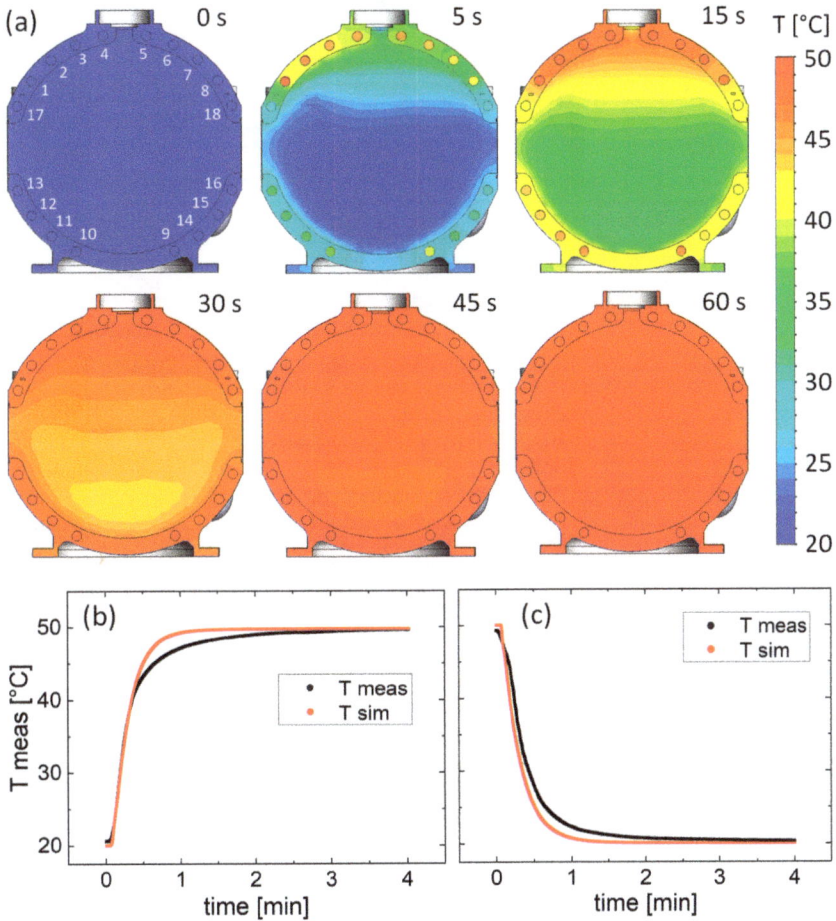

Figure 2. (a) Contour plots across the sectional views of the chamber during the heating simulation over a period of one minute. Each color corresponds to an isothermal range of 1 °C. The numbers in the first image indicate the flow order of the liquid flow. Temperature was measured at the center of the chamber during the simulation (red) and the laboratory experiment (black) for (b) heating from 20 to 50 °C and (c) cooling from 50 to 20 °C.

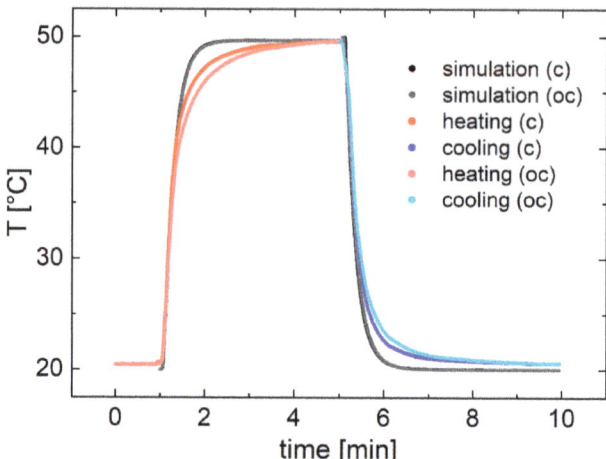

Figure 3. Temperature sensor data and simulated temperature during the heating and cooling process at the center (c) of the chamber and 35 mm off-center (oc) in the direction of the windows.

2.2. Gas Flow Setup

In order to control and regulate the vapor composition within the chamber, a gas flow setup was installed on top of the optical breadboard. The layout of the gas flow is depicted in Figure 4 and can be grouped into the gas flow circuit (orange), an electrical control circuit (green), and a temperature control circuit (blue). The gas flow circuit consists of a supply of dry nitrogen (A), three remote-controllable gas flow controllers (B), two washing bottles (C), the connection to the humidity chamber (D), and, finally, the gas outlet (E). The gas flow controllers (B) are also part of the electrical control circuit, together with the power supply, the readout system (F), and the required software running on a computer (G). Lastly, the temperature control circuit consists of a thermal bath (H), the tubing within the gas flow chamber, the channels within the humidity chamber, and additional tubing to connect the various parts.

The core parts of the control circuit are three F-201CV-1K0 gas flow controllers calibrated with a system pressure of 2 bar at 20 °C to adjust a nitrogen flow rate between 0 and 1 L min^{-1}, as well as the E-8501-R-00 supply and readout system that provide the power supply and handle the readout and control of the three gas flow controllers (both by Bronkhorst High-Tech B.V.Ak Ruurlo, Netherlands). They are connected by a set of patch cables and a multiport adapter with one connection leading to a remotely located computer via an RS232 connection. The PC is equipped with the FlowDDE software, which is a dynamic data exchange (DDE) server that connects digital instruments to windows applications, such as the FlowView software, a DDE-client program that is used to operate digital instruments. With the software, the gas flow of all three controllers can be adjusted and monitored.

The three flow controllers provide three individual and steady nitrogen gas flows. Two of them are led via polytetrafluoroethylene (PTFE) tubing to two separate 200 mL washing bottles (Neubert-Glas GbR, Geschwenda, Germany) equipped with a glass frit, while the third gas flow is a pure nitrogen line used for drying the chamber or to generate intermediate humidity values. The washing bottles can be filled with water or other solvents, thus allowing for different pure or mixed solvent atmospheres within the chamber. In order to prevent back flow of any solvated gas to the controllers and any uncontrolled diffusion from the washing bottles into the gas supply to the chamber, non-return valves are arranged before and after the washing bottles, as well as on the pure nitrogen line. The three gas streams merge into one and are led via a water-heated tube from the gas flow chamber to the humidity chamber. The gas flow enters the chamber below the aluminum plate on which the sample holder pins are located. Then it diffuses up to the sample through a set of holes in the sample holder base plate in order to prevent turbulences and atmospheric inhomogeneity around the sample. Finally, as the chamber is airtight, an exhaust tube leads the gas flow through another washing bottle to clean the gas flow from potentially hazardous substances.

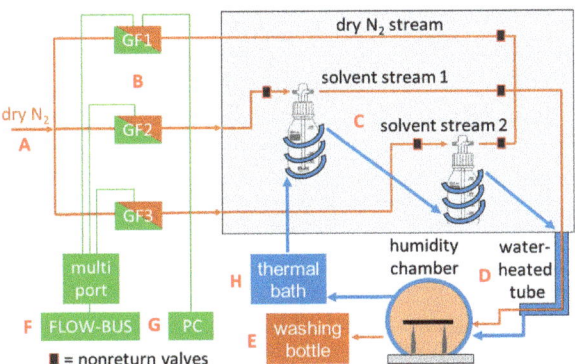

Figure 4. Scheme of the gas flow system. The gas flow circuit (orange) consists of a dry nitrogen supply (**A**) regulated by three gas flow controllers (**B**), two washing bottles for humidification (**C**), a water heated tube to the chamber (**D**), and a washing bottle for gas cleaning (**E**). The control circuit (green) consists of the gas flow controllers (**B**), a power supply and readout system (**F**), and a computer with the running software (**G**). The temperature circuit (blue) consists of a thermal bath (**H**) and the tubing around the washing bottles to the chamber and within the chamber itself. Non-return valves are used to prevent backflow and uncontrolled diffusion from the washing bottles.

The temperature of the whole system is regulated by a single thermal bath. As it flows through copper tubes coiled around the bottles, it heats the washing bottles and the whole gas flow chamber. The gas flow chamber is insulated and equipped with a computer fan to ensure a homogeneous temperature distribution, which is monitored with an SHT31 sensor. The heated liquid then flows to the humidity chamber via silicone tubes coiled around the gas flow tube, keeping the gas temperature constant. After heating the chamber as described in the previous section, the liquid flows back to the thermal bath.

In order to test the gas flow system, we adjusted the gas flow through one water-filled washing bottle from 0 to 100% (100% was the equivalent to 1 L min^{-1}) in 20% steps and adjusted the gas flow on the pure nitrogen line accordingly to a total flow of 1 L min^{-1} (N_2 flow at 100 down to 0%). Then, we increased the temperature from 20 to 50 °C and repeated the same experiment. The evolution of the temperature and humidity at the chambers center were recorded with an SHT31 sensor and are shown in Figure 5. The humidity within the chamber starts at 0% RH (0% H_2O and 100% N_2 flow) before rising to 20% over the course of around 7 min as we adjust the flow rates accordingly (20% H_2O and 80% N_2 flow). The measured humidity continues to follow the adjusted flow rates in 20% steps, which proves the good control of the gas flow system. Eventually, a humidity of 96%RH (100% H_2O and 0% N_2 flow) is reached. Condensational losses at the windows and the tubes prevented the humidity from being even higher, although no visible evidence of condensation during and after any performed experiments was found. Furthermore, the equilibration time increases as we aim to obtain a pure water or nitrogen atmosphere, which is due to the relatively large volume of the chamber. After adjusting the flow rates back to a pure nitrogen flow, it takes around 17 min to dry the atmosphere back to below 1% RH. The repeated experiment at a temperature of 50 °C, shows an equally good control over the humidity and similar equilibration times. The highest achieved humidity is slightly lower at 93% RH due to the stronger temperature gradient at the windows. In conclusion, the presented gas flow system allows for the study of a broad and well-controlled temperature and humidity range despite the large chamber volume and generous window sizes. This system was already used in neutron scattering experiments that investigated thin-film responses under changing atmospheric compositions, which also demonstrated the use of solvents other than water [45,46]. The setup currently runs without a feedback loop, as set humidity (via the flow rates) and temperature values are reached accurately.

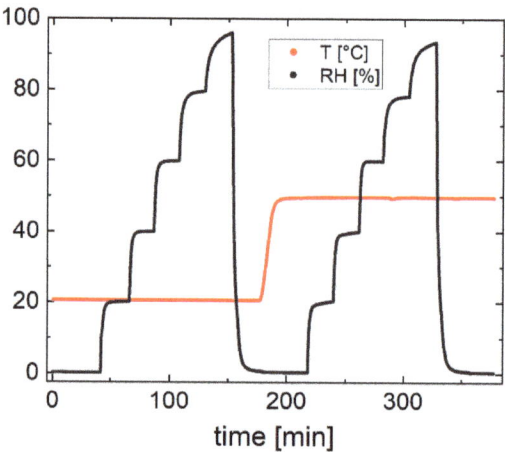

Figure 5. Humidity (black) and temperature (red) evolution in the center of the humidity chamber during two step-wise increases of the flow rate of a water saturated gas flow mixed with a pure nitrogen flow at 20 and 50 °C. The flow rate of the water-saturated flow started at 0% and was increased in 20% steps to 100%, while the flow rate of the nitrogen stream started at 100% and was decreased accordingly to 0%—thus keeping the total flow at 100%, which was equivalent to a constant flow of 1 L min^{-1}. After each humidification cycle, the flow rates were switched back to 0% water and 100% nitrogen, drying the atmosphere. The highest reached humidity is 96% RH (20 °C) and 93% RH (50 °C).

2.3. In Situ Spectral Reflectance Setup

To complement the neutron scattering data on thin films, thickness measurement techniques are commonly applied, which enable the correlation of morphological and compositional changes within the film to the changes in film thickness. In order to realize this in situ, a spectral reflectance setup was connected to the lid of the humidity chamber. Spectral reflectance can probe film thickness in the nanometer-to-micrometer range and is well-suited to the typical film thicknesses investigated with neutron scattering techniques. Additionally, it provides information on the refractive index of a film, which contributes to the determination of film porosity or water content. Figure 6a schematically shows the spectral reflectance setup. Light from a light source (I) is directed through an optical fiber to a lens assembly (II) on top of the chamber lid. The light is reflected from the sample surface (III), and the thin film interference signal is detected by a spectrometer (IV) and analyzed on an external computer (V).

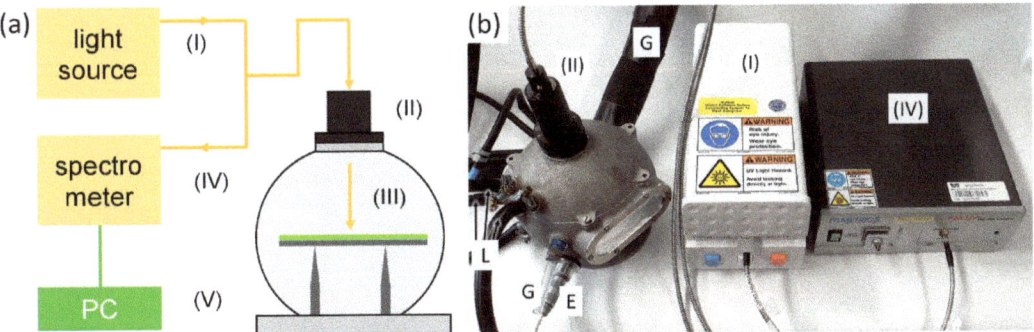

Figure 6. (**a**) Scheme of the spectral reflectance setup consisting of a light source (I), a lens assembly to focus the light (II), the sample on which the light gets reflected (III), the spectrometer recording the reflected light (IV), and a computer including the control and analysis software (V). (**b**) Photograph of the assembly. Connections to the chamber are labeled for the liquid flow (L), the gas flow (G), and the electronics (E) connecting to the sensors.

We use an LS-DT2 deuterium-tungsten light source with a remote-controlled shutter, together with the F20-UV thin film analyzer spectrometer and the LA-GL25-25-25-UV lens assembly in the KM-GL25 kinematic mount (all by Filmmetrics Europe GmbH, Unterhaching, Germany), which is fixed to the extrusion on the chamber lid (see Figure 1c). The LS-DT2 provides a wavelength band from 190 to 1050 nm and is able to probe films within a film thickness range from 1 nm to 40 µm. The one-inch UV lens with a focal length of 25.4 mm is positioned within the lens assembly at a distance of 103 mm above the sample position. A calcium fluoride window installed in the lid of the chamber allows for light in the UV/Vis and NIR range to pass through a 1-cm-diameter hole in to chamber lid. Prior to measurements, a reference measurement of the pristine substrate reflectance is recorded and subtracted from the sample data. Recorded spectra are analyzed with the FILMeasure software using a vertical layer model and give live feedback of the thickness and refractive index during the neutron experiment independently of the collected neutron data.

For exemplary measurements, we prepared a thin film of a poly(N-isopropylacrylamide) (PNIPAM) microgel cross-linked with N,N'-methylenebisacrylamide on a silicon substrate. PNIPAM microgels are temperature-responsive and highly hygroscopic at temperatures below the volume phase transition temperature (VPPT) [47–49]. The film was placed in the humidity chamber at a temperature of 20 °C (well below the VPPT [38]) and humidified in 20% RH steps from a dry atmosphere at 0% RH to a humidity of 96% RH. The spectral reflectance setup continuously recorded a spectrum every 10 s. A simple two-layer model, SiO_2 and a generic polymer layer with a refractive index of 1.45, was used to model-fit the spectra. The extracted film thickness is shown in Figure 7 as a function of measurement time, together with the humidity within the chamber.

The obtained thickness data correlates well with the humidity within the chamber. A good fit (goodness of fit >0.999) was achieved for all collected spectra, and a continuous thickness increase of the film with rising humidity is observed. A stronger swelling at higher humidity is observed and is in accordance with neutron reflectometry data collected on such PNIPAM microgel thin films [38]. Overall, we are able to accurately track the thickness via the spectral reflectance setup attached to the chamber lid, allowing for complementary in situ spectral reflectance measurements during GISANS or NR measurements. The setup was already used during a recent NR study and proved to be an important tool to measure the film thickness and refractive index in situ [46].

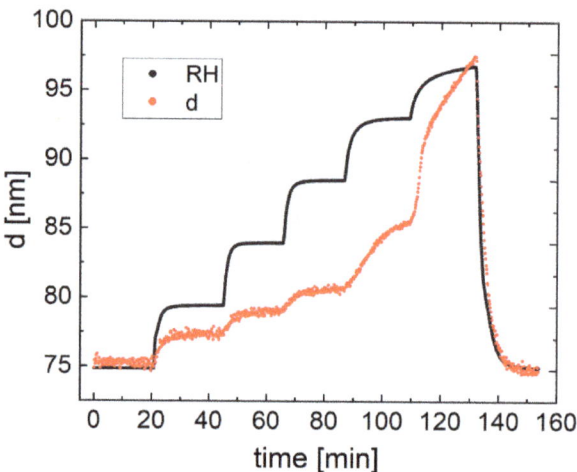

Figure 7. Film thickness (red) and humidity (black) during a continuous spectral reflectance measurement at 20 °C and a 20%RH step-wise increase of the humidity from 0% to 96% RH.

2.4. Motion Control and Sample Alignment

The alignment of thin film samples for GISANS or NR measurements is crucial and requires high accuracy. Therefore, an inherent alignment system was installed on top of the optical breadboard consisting of a lateral y-translation stage perpendicular to the neutron beam (set as x direction hereafter), a vertical z-translation stage, and an angular translation stage for both rotations around the x- and y-axes. For the y-translation stage, the LS-180 (PI miCos GmbH, Eschbach, Germany) with a 508 mm travel range (as seen in Figure 8a,b) is used. It is equipped with a two-phase bipolar half-coil stepper motor (PK-258-02B, Oriental Motor) and a linear optical encoder with an RS422 quadrature output signal (LIA 20, NUMERIK JENA), and it allows for the positioning of an up-to 100 kg load with an accuracy of 1 μm. The z-translation is achieved with the 5103.A10-35 z-stage, and the angular translation is achieved with the 5203.10 2-circle-segment (both Huber Diffraktionstechnik GmbH & Co. KG, Rimsting, Germany). They are shown in Figure 8c,d equipped with two-phase hybrid stepper motors (ZSS43.200.1.2, Phytron) and optical absolute rotary encoders (OCD-S101G-1413-S060-PRL, Posital/Fraba) with an accuracy of 0.022° (multiturn). The z-stage offers a travel range of 35 mm, while the 2-circle segment offers an angular range of ±14°, which even covers strong tilts as well as the usual angles of up to around 3° that are used in GISANS or NR. Initial height adjustments with a 0.1 mm accuracy are possible by the lifting table that is part of the mounting system. The stages are assembled as shown in Figure 8e, with the linear stage screwed to the optical breadboard, the z-stage mounted on top of the linear stage, and the 2-circle segment located at the very top. The pivot point of the 2-circle segment is 110 mm above its top edge, and the position of the sample within the humidity chamber is 74 mm above its bottom edge. As such, a 36 mm spacer with a 10 mm PTFE plate is used between the 2-circle segment and the chamber, which additionally functions as insulation of the alignment components from the temperature changes of the chamber.

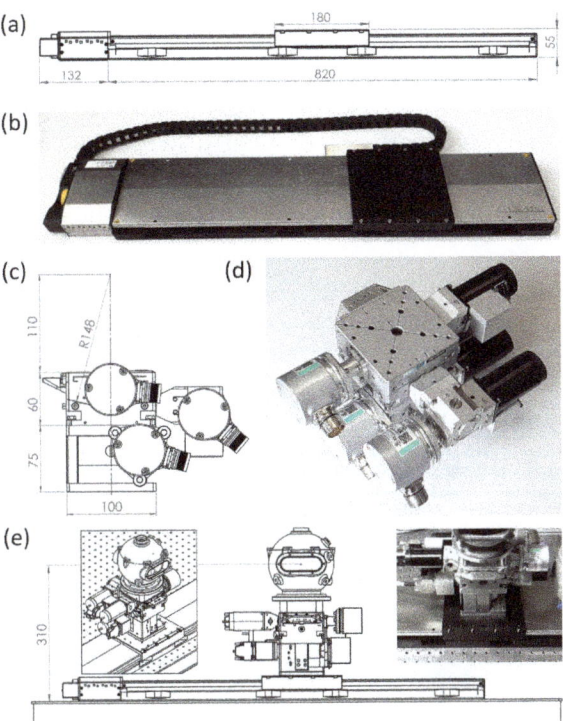

Figure 8. (**a**) Technical drawing and (**b**) photograph of the LS180, used for the horizontal sample alignment. (**c**) Technical drawing and (**d**) photograph of the 5103.A10-35 and the 5203.10, used for the vertical and rotational alignment, respectively. (**e**) Assembly of the components on the breadboard.

In order to control the different stages, a custom-made control unit was assembled based on the Ethernet-based fieldbus system EtherCAT [50] (Ethernet for Control Automation Technology) (Beckhoff Automation GmbH & Co. KG, Verl, Germany). The layout of the motion control unit is shown in Figure 9a. The motors (M1–4) and encoders (E1–4) belong to the LS-180 linear stage (M1, E1), the z-stage (M2, E2), and the 2-circle segment (M3, M4, E3, E4). Three power supply units are installed in a 19-inch rack together with an embedded computer and the required EtherCAT terminals. The 48 V supply feeds the 4 motors, while the 24 V power lines feeds the CX5130 embedded PC and thereby also the internal terminal power circuit. The EL9189 is a 0 V potential distribution terminal to which M1 was connected. The two EL1808 24 V are digital input terminals that registered the limit switches of the four motors (two per terminal). The EL2819 is a digital output terminal that connects the 24 V signal of the automation unit to the three encoders of the z-stage and 2-circle segment, and the two EL5002 are the SSI encoder interfaces for these three encoders. Since E1 provides an RS422 signal, EL5101, which is an incremental encoder interface that registers RS422 differential signals, is used. The four EL7041-0052 are stepper motor terminals to which M1–M4 are directly connected. Lastly, the EL9011 is a bus end cover that terminated the bus station and covers the contacts. If not supplied via a terminal, the motors, encoders, and cable shields are grounded via the power supply.

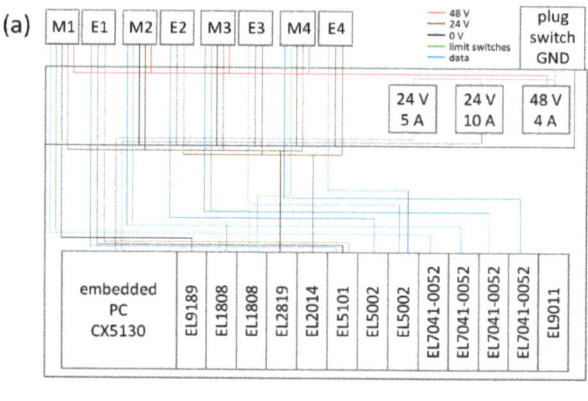

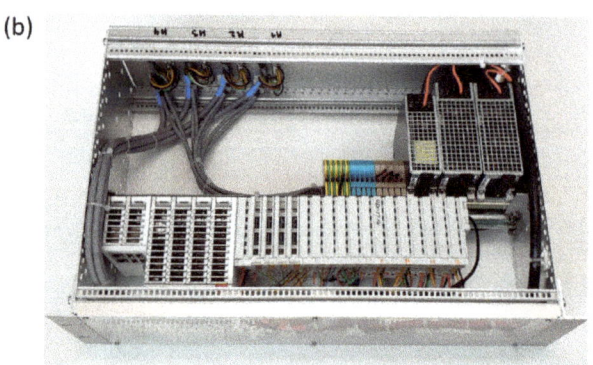

Figure 9. (a) Wiring of the motion control unit for the four axes. M1–4 and E1–4 correspond to the motors and encoders of the 4 axes (LS-180 is axis 1). The EtherCAT terminals (ELXXXX) control the power supply and data flow of the encoders and motors. (b) Photograph of the assembled control unit.

The complete assembly of the unit can be seen in the photograph shown in Figure 9b, with all cables connected. In order to access and control the different axes, TwinCAT 3 is installed on the embedded computer. In cooperation with the motion control group at the ESS, the motors and encoders of the used stages were integrated into a project that allows for the accurate control over each axis and their remote access. Laboratory tests were conducted to ensure that all axes are working properly.

3. Conclusions

In the framework of the FlexiProb project, we designed and built a sample environment for the investigation of soft matter thin films via GISANS for the use at the ESS. The presented setup includes four core components: a spherical, 3D printed humidity chamber; a gas flow control setup; a spectral reflectance measurement setup; and a motion control system. The 3D-printed spherical humidity chamber with embedded fluid channels is optimized to reduce condensation and provide a homogeneous temperature distribution around the thin film sample. Temperature simulations and measurements show a layered temperature profile for temperature switches, with no significant temperature gradient even over large samples. The three-channel gas flow setup provides well-controlled mixed vapor atmospheres of pure nitrogen and two solvents. Gas compositions within the chamber can be easily adjusted via the regulation of the flow rates of the three channels. Laboratory experiments confirmed the accurate control and a humidity of 96%RH was achieved for an exemplary water saturated gas-stream despite the large chamber volume. The spectral reflectance measurement setup installed on top of the chamber lid enables in situ thickness measurements during neutron experiments. Exemplary measurements demonstrated the functionality of the setup. Finally, the motion control setup consisting of four axes for lateral (perpendicular to neutron beam), vertical, and angular movement enables the accurate positioning of the sample for alignment and experimentation. The custom assembled control unit is based on the Beckhoff EtherCAT technology and optimized for future integration at the ESS. Thus, for the ESS instruments, a versatile GISANS sample environment that can be used for early science at the ESS is available.

Author Contributions: Conceptualization, R.v.K., T.H., and P.M.-B.; planning and discussion of setup components, T.W., L.P.K., M.K., A.J.S., L.W., S.J., H.F., O.L., H.S., and A.H.; construction and test of setup components, T.W. and L.P.K.; writing—original draft preparation, T.W.; writing—review and editing, L.P.K., M.K., A.J.S., L.W., S.J., H.F., O.L., H.S., A.H., R.v.K., T.H., and P.M.-B.; visualization, T.W. and L.P.K.; supervision, P.M.-B.; project administration, T.H.; funding acquisition, R.v.K., T.H., and P.M.-B. All authors have read and agreed to the published version of the manuscript.

Funding: This research was funded by the German Federal Ministry for Education and Research (BMBF) within the project "FlexiProb" sample environment grant number 05K2016.

Institutional Review Board Statement: Not applicable.

Informed Consent Statement: Not applicable.

Data Availability Statement: Derived data supporting the findings of this study are available from the corresponding author upon reasonable request.

Acknowledgments: We gratefully acknowledge the ESS Motion Control Group for help implementing the control of the translation stages. We thank Judith Houston for fruitful discussions. We thank Christina Geiger and Julija Reitenbach for valuable discussions and testing of setup components.

Conflicts of Interest: The authors declare no conflict of interest.

References

1. Kiel, J.W.; Mackay, M.E.; Kirby, B.J.; Maranville, B.B.; Majkrzak, C.F. Phase-sensitive neutron reflectometry measurements applied in the study of photovoltaic films. *J. Chem. Phys.* **2010**, *133*, 74902. [CrossRef]
2. Clulow, A.J.; Tao, C.; Lee, K.H.; Velusamy, M.; McEwan, J.A.; Shaw, P.E.; Yamada, N.L.; James, M.; Burn, P.L.; Gentle, I.R.; et al. Time-resolved neutron reflectometry and photovoltaic device studies on sequentially deposited PCDTBT-fullerene layers. *Langmuir* **2014**, *30*, 11474–11484. [CrossRef]
3. Jerliu, B.; Dörrer, L.; Hüger, E.; Borchardt, G.; Steitz, R.; Geckle, U.; Oberst, V.; Bruns, M.; Schneider, O.; Schmidt, H. Neutron reflectometry studies on the lithiation of amorphous silicon electrodes in lithium-ion batteries. *Phys. Chem. Chem. Phys.* **2013**, *15*, 7777–7784. [CrossRef]
4. Ohisa, S.; Matsuba, G.; Yamada, N.L.; Pu, Y.-J.; Sasabe, H.; Kido, J. Precise Evaluation of Angstrom-Ordered Mixed Interfaces in Solution-Processed OLEDs by Neutron Reflectometry. *Adv. Mater. Interfaces* **2014**, *1*, 1400097. [CrossRef]
5. Ueda, S.; Koizumi, S.; Ohira, A.; Kuroda, S.; Frielinghaus, H. Grazing-incident neutron scattering to access catalyst for polymer electrolyte fuel cell. *Physica B* **2018**, *551*, 309–314. [CrossRef]
6. Cavaye, H.; Smith, A.R.G.; James, M.; Nelson, A.; Burn, P.L.; Gentle, I.R.; Lo, S.-C.; Meredith, P. Solid-state dendrimer sensors: Probing the diffusion of an explosive analogue using neutron reflectometry. *Langmuir* **2009**, *25*, 12800–12805. [CrossRef] [PubMed]
7. Junghans, A.; Watkins, E.B.; Barker, R.D.; Singh, S.; Waltman, M.J.; Smith, H.L.; Pocivavsek, L.; Majewski, J. Analysis of biosurfaces by neutron reflectometry: From simple to complex interfaces. *Biointerphases* **2015**, *10*, 19014. [CrossRef]
8. Simič, R.; Kalin, M.; Hirayama, T.; Korelis, P.; Geue, T. Fatty Acid Adsorption on Several DLC Coatings Studied by Neutron Reflectometry. *Tribol. Lett.* **2014**, *53*, 199–206. [CrossRef]
9. Cousin, F.; Jestin, J.; Chaboussant, G.; Gautrot, S.; Menelle, A.; Ott, F. Probing simultaneously the volume and surface structure of nanospheres adsorbed at a solid-liquid interface by GISANS. *Eur. Phys. J. Spec. Top.* **2009**, *167*, 177–183. [CrossRef]
10. Lipfert, F.; Frielinghaus, H.; Holderer, O.; Mattauch, S.; Monkenbusch, M.; Arend, N.; Richter, D. Polymer enrichment decelerates surfactant membranes near interfaces. *Phys. Rev. E Stat. Nonlinear Soft Matter Phys.* **2014**, *89*, 42303. [CrossRef]
11. Russell, T.P. X-ray and neutron reflectivity for the investigation of polymers. *Mater. Sci. Rep.* **1990**, *5*, 171–271. [CrossRef]
12. Penfold, J.; Thomas, R.K. The application of the specular reflection of neutrons to the study of surfaces and interfaces. *J. Phys. Condens. Matter* **1990**, *2*, 1369–1412. [CrossRef]
13. Majkrzak, C.F.; Felcher, G.P. Neutron Scattering Studies of Surfaces and Interfaces. *MRS Bull.* **1990**, *15*, 65–72. [CrossRef]
14. Müller-Buschbaum, P.; Gutmann, J.S.; Stamm, M. Dewetting of confined polymer films: An X-ray and neutron scattering study. *Phys. Chem. Chem. Phys.* **1999**, *1*, 3857–3863. [CrossRef]

15. Müller-Buschbaum, P.; Gutmann, J.S.; Cubitt, R.; Stamm, M. Probing the in-plane composition of thin polymer films with grazing-incidence small-angle neutron scattering and atomic force microscopy. *Colloid Polym. Sci.* **1999**, *277*, 1193–1199. [CrossRef]
16. Kraus, J.; Müller-Buschbaum, P.; Kuhlmann, T.; Schubert, D.W.; Stamm, M. Confinement effects on the chain conformation in thin polymer films. *Eur. Lett.* **2000**, *49*, 210–216. [CrossRef]
17. Helmholtz Zentrum Berlin Website. Available online: https://www.helmholtz-berlin.de/projects/rueckbau/ber/index_en.html (accessed on 28 April 2021).
18. Laboratoire Léon Brillouin Website. Available online: http://www-llb.cea.fr/en/Phocea/Vie_des_labos/News/index.php?id_news=7680 (accessed on 28 April 2021).
19. Institute for Energy Technology (IFE) Website. Available online: https://ife.no/en/permanent-closure-of-the-jeep-ii-research-reactor-at-kjeller/ (accessed on 28 April 2021).
20. Garoby, R.; Vergara, A.; Danared, H.; Alonso, I.; Bargallo, E.; Cheymol, B.; Darve, C.; Eshraqi, M.; Hassanzadegan, H.; Jansson, A.; et al. The European Spallation Source Design. *Phys. Scr.* **2018**, *93*, 14001. [CrossRef]
21. Andersen, K.H.; Argyriou, D.N.; Jackson, A.J.; Houston, J.; Henry, P.F.; Deen, P.P.; Toft-Petersen, R.; Beran, P.; Strobl, M.; Arnold, T.; et al. The instrument suite of the European Spallation Source. *Nucl. Instrum. Methods Phys. Res. Sect. A* **2020**, *957*, 163402. [CrossRef]
22. Jackson, A.; Kanaki, K. Ess Construction Proposal: Loki—A Broad-Band Sans Instrument. *Zenodo* **2013**. [CrossRef]
23. Jaksch, S.; Chennevière, A.; Désert, S.; Kozielewski, T.; Feilbach, H.; Lavie, P.; Hanslik, R.; Gussen, A.; Butterweck, S.; Engels, R.; et al. Technical Specification of the Small-Angle Neutron Scattering Instrument SKADI at the European Spallation Source. *Applied Sciences* **2021**, *11*, 3620. [CrossRef]
24. Frielinghaus, H.; Gvaramia, M.; Mangiapia, G.; Jaksch, S.; Ganeva, M.; Koutsioubas, A.; Mattauch, S.; Ohl, M.; Monkenbusch, M.; Holderer, O. New tools for grazing incidence neutron scattering experiments open perspectives to study nano-scale tribology mechanisms. *Nucl. Instrum. Methods Phys. Res. Sect. A* **2017**, *871*, 72–76. [CrossRef]
25. Schmidt, S.; Motschmann, H.; Hellweg, T.; von Klitzing, R. Thermoresponsive surfaces by spin-coating of PNIPAM-co-PAA microgels: A combined AFM and ellipsometry study. *Polymer* **2008**, *49*, 749–756. [CrossRef]
26. Hertle, Y.; Hellweg, T. Thermoresponsive copolymer microgels. *J. Mater. Chem. B* **2013**, *1*, 5874. [CrossRef]
27. Kreuzer, L.P.; Widmann, T.; Bießmann, L.; Hohn, N.; Pantle, J.; Märkl, R.; Moulin, J.-F.; Hildebrand, V.; Laschewsky, A.; Papadakis, C.M.; et al. Phase Transition Kinetics of Doubly Thermoresponsive Poly(sulfobetaine)-Based Diblock Copolymer Thin Films. *Macromolecules* **2020**, *53*, 2841–2855. [CrossRef]
28. Kreuzer, L.P.; Widmann, T.; Hohn, N.; Wang, K.; Bießmann, L.; Peis, L.; Moulin, J.-F.; Hildebrand, V.; Laschewsky, A.; Papadakis, C.M.; et al. Swelling and Exchange Behavior of Poly(sulfobetaine)-Based Block Copolymer Thin Films. *Macromolecules* **2019**, *52*, 3486–3498. [CrossRef]
29. Wang, W.; Kaune, G.; Perlich, J.; Papadakis, C.M.; Bivigou Koumba, A.M.; Laschewsky, A.; Schlage, K.; Röhlsberger, R.; Roth, S.V.; Cubitt, R.; et al. Swelling and switching kinetics of gold coated end-capped poly(N-isopropylacrylamide) thin films. *Macromolecules* **2010**, *43*, 2444–2452. [CrossRef]
30. Gonthier, J.; Barrett, M.A.; Aguettaz, O.; Baudoin, S.; Bourgeat-Lami, E.; Demé, B.; Grimm, N.; Hauß, T.; Kiefer, K.; Lelièvre-Berna, E.; et al. BerILL: The ultimate humidity chamber for neutron scattering. *J. Neutron Res.* **2019**, *21*, 65–76. [CrossRef]
31. Plaza, N.Z.; Pingali, S.V.; Qian, S.; Heller, W.T.; Jakes, J.E. Informing the improvement of forest products durability using small angle neutron scattering. *Cellulose* **2016**, *23*, 1593–1607. [CrossRef]
32. Arima-Osonoi, H.; Miyata, N.; Yoshida, T.; Kasai, S.; Ohuchi, K.; Zhang, S.; Miyazaki, T.; Aoki, H. Gas-flow humidity control system for neutron reflectivity measurements. *Rev. Sci. Instrum.* **2020**, *91*, 104103. [CrossRef]
33. Carotenuto, A.; Dell'Isola, M. An experimental verification of saturated salt solution-based humidity fixed points. *Int. J. Thermophys.* **1996**, *17*, 1423–1439. [CrossRef]
34. Harroun, T.A.; Fritzsche, H.; Watson, M.J.; Yager, K.G.; Tanchak, O.M.; Barrett, C.J.; Katsaras, J. Variable temperature, relative humidity (0–100%), and liquid neutron reflectometry sample cell suitable for polymeric and biomimetic materials. *Rev. Sci. Instrum.* **2005**, *76*, 65101. [CrossRef]
35. Young, J.F. Humidity control in the laboratory using salt solutions-a review. *J. Appl. Chem.* **1967**, *17*, 241–245. [CrossRef]
36. Demé, B.; Cataye, C.; Block, M.A.; Maréchal, E.; Jouhet, J. Contribution of galactoglycerolipids to the 3-dimensional architecture of thylakoids. *FASEB J.* **2014**, *28*, 3373–3383. [CrossRef] [PubMed]
37. Bießmann, L.; Kreuzer, L.P.; Widmann, T.; Hohn, N.; Moulin, J.-F.; Müller-Buschbaum, P. Monitoring the Swelling Behavior of PEDOT:PSS Electrodes under High Humidity Conditions. *ACS Appl. Mater. Interfaces* **2018**, *10*, 9865–9872. [CrossRef] [PubMed]
38. Widmann, T.; Kreuzer, L.P.; Hohn, N.; Bießmann, L.; Wang, K.; Rinner, S.; Moulin, J.-F.; Schmid, A.J.; Hannappel, Y.; Wrede, O.; et al. Hydration and Solvent Exchange Induced Swelling and Deswelling of Homogeneous Poly(N-isopropylacrylamide) Microgel Thin Films. *Langmuir* **2019**, *35*, 16341–16352. [CrossRef] [PubMed]
39. Thijs, H.M.L.; Becer, C.R.; Guerrero-Sanchez, C.; Fournier, D.; Hoogenboom, R.; Schubert, U.S. Water uptake of hydrophilic polymers determined by a thermal gravimetric analyzer with a controlled humidity chamber. *J. Mater. Chem.* **2007**, *17*, 4864. [CrossRef]
40. Oerter, E.J.; Singleton, M.; Thaw, M.; Davisson, M.L. Water vapor exposure chamber for constant humidity and hydrogen and oxygen stable isotope composition. *Rapid Commun. Mass Spectrom.* **2019**, *33*, 89–96. [CrossRef]

41. Feoktystov, A.V.; Frielinghaus, H.; Di, Z.; Jaksch, S.; Pipich, V.; Appavou, M.-S.; Babcock, E.; Hanslik, R.; Engels, R.; Kemmerling, G.; et al. KWS-1 high-resolution small-angle neutron scattering instrument at JCNS: Current state. *J. Appl. Crystallogr.* **2015**, *48*, 61–70. [CrossRef]
42. Frielinghaus, H.; Feoktystov, A.; Berts, I.; Mangiapia, G. KWS-1: Small-angle scattering diffractometer. *JLSRF* **2015**, *1*. [CrossRef]
43. Widmann, T.; Kreuzer, L.P.; Mangiapia, G.; Haese, M.; Frielinghaus, H.; Müller-Buschbaum, P. 3D printed spherical environmental chamber for neutron reflectometry and grazing-incidence small-angle neutron scattering experiments. *Rev. Sci. Instrum.* **2020**, *91*, 113903. [CrossRef]
44. Medici, M.-G.; Mongruel, A.; Royon, L.; Beysens, D. Edge effects on water droplet condensation. *Phys. Rev. E: Stat. Nonlinear Soft Matter Phys.* **2014**, *90*, 62403. [CrossRef] [PubMed]
45. Kreuzer, L.P.; Lindenmeir, C.; Geiger, C.; Widmann, T.; Hildebrand, V.; Laschewsky, A.; Papadakis, C.M.; Müller-Buschbaum, P. Poly(sulfobetaine) versus Poly(N-isopropylmethacrylamide): Co-Nonsolvency-Type Behavior of Thin Films in a Water/Methanol Atmosphere. *Macromolecules* **2021**, *54*, 1548–1556. [CrossRef]
46. Geiger, C.; Reitenbach, J.; Kreuzer, L.P.; Widmann, T.; Wang, P.; Cubitt, R.; Henschel, C.; Laschewsky, A.; Papadakis, C.M.; Müller-Buschbaum, P. PMMA-b-PNIPAM Thin Films Display Cononsolvency-Driven Response in Mixed Water/Methanol Vapors. *Macromolecules* **2021**, *54*, 3517–3530. [CrossRef]
47. Burmistrova, A.; Richter, M.; Eisele, M.; Üzüm, C.; Klitzing, R. von. The Effect of Co-Monomer Content on the Swelling/Shrinking and Mechanical Behaviour of Individually Adsorbed PNIPAM Microgel Particles. *Polymers* **2011**, *3*, 1575–1590. [CrossRef]
48. Burmistrova, A.; von Klitzing, R. Control of number density and swelling/shrinking behavior of P(NIPAM–AAc) particles at solid surfaces. *J. Mater. Chem.* **2010**, *20*, 3502–3507. [CrossRef]
49. Backes, S.; Krause, P.; Tabaka, W.; Witt, M.U.; von Klitzing, R. Combined Cononsolvency and Temperature Effects on Adsorbed PNIPAM Microgels. *Langmuir* **2017**, *33*, 14269–14277. [CrossRef] [PubMed]
50. Buttner, H.; Jansen, D. Real-time Ethernet: The EtherCAT solution. *Comput. Control Eng.* **2004**, *15*, 16–21. [CrossRef]

Article

Flexible Sample Environments for the Investigation of Soft Matter at the European Spallation Source: Part III—The Macroscopic Foam Cell

Matthias Kühnhammer [1], Tobias Widmann [2], Lucas P. Kreuzer [2], Andreas J. Schmid [3], Lars Wiehemeier [3], Henrich Frielinghaus [4], Sebastian Jaksch [4], Torsten Bögershausen [5], Paul Barron [5], Harald Schneider [5], Arno Hiess [5], Peter Müller-Buschbaum [2,6], Thomas Hellweg [3], Regine von Klitzing [1,*] and Oliver Löhmann [1,5,†]

1 Institut für Physik Kondensierter Materie, Technische Universität Darmstadt, Hochschulstraße 8, 64289 Darmstadt, Germany; kuehnhammer@fkp.tu-darmstadt.de (M.K.); oliver.loehmann@bam.de (O.L.)
2 Lehrstuhl für Funktionelle Materialien, Physik Department, Technische Universität München, James-Franck-Str. 1, 85748 Garching, Germany; Tobias.widmann@ph.tum.de (T.W.); lucas.kreuzer@ph.tum.de (L.P.K.); muellerb@ph.tum.de (P.M.-B.)
3 Fakultät für Chemie, Physikalische und Biophysikalische Chemie, Universität Bielefeld, Universitätsstr. 25, 33615 Bielefeld, Germany; andreas.josef.schmid@rwth-aachen.de (A.J.S.); lars.wiehemeier@uni-bielefeld.de (L.W.); thomas.hellweg@uni-bielefeld.de (T.H.)
4 Jülich Centre for Neutron Science JCNS at Heinz Maier-Leibnitz Zentrum (MLZ), Forschungszentrum Jülich GmbH, Lichtenbergstr, 85748 Garching, Germany; h.frielinghaus@fz-juelich.de (H.F.); s.jaksch@fz-juelich.de (S.J.)
5 European Spallation Source ERIC, P.O. Box 176, SE-221 00 Lund, Sweden; torsten.bogershausen@esss.se (T.B.); Paul.Barron@esss.se (P.B.); Harald.Schneider@ess.eu (H.S.); arno.hiess@ess.eu (A.H.)
6 Heinz Maier-Leibnitz Zentrum (MLZ), Technische Universität München, Lichtenbergstr. 1, 85748 Garching, Germany
* Correspondence: klitzing@smi.tu-darmstadt.de
† Current address: Bundesanstalt für Materialforschung Und-Prüfung (BAM), Unter den Eichen 87, 12205 Berlin, Germany.

Citation: Kühnhammer, M.; Widmann, T.; Kreuzer, L.P.; Schmid, A.J.; Wiehemeier, L.; Frielinghaus, H.; Jaksch, S.; Bögershausen, T.; Barron, P.; Schneider, H.; et al. Flexible Sample Environments for the Investigation of Soft Matter at the European Spallation Source: Part III—The Macroscopic Foam Cell. *Appl. Sci.* **2021**, *11*, 5116. https://doi.org/10.3390/app11115116

Academic Editor: Antonino Pietropaolo

Received: 30 April 2021
Accepted: 25 May 2021
Published: 31 May 2021

Publisher's Note: MDPI stays neutral with regard to jurisdictional claims in published maps and institutional affiliations.

Copyright: © 2021 by the authors. Licensee MDPI, Basel, Switzerland. This article is an open access article distributed under the terms and conditions of the Creative Commons Attribution (CC BY) license (https://creativecommons.org/licenses/by/4.0/).

Abstract: The European Spallation Source (ESS), which is under construction in Lund (Sweden), will be the leading and most brilliant neutron source and aims at starting user operation at the end of 2023. Among others, two small angle neutron scattering (SANS) machines will be operated. Due to the high brilliance of the source, it is important to minimize the downtime of the instruments. For this, a collaboration between three German universities and the ESS was initialized to develop and construct a unified sample environment (SE) system. The main focus was set on the use of a robust carrier system for the different SEs, which allows setting up experiments and first prealignment outside the SANS instruments. This article covers the development and construction of a SE for SANS experiments with foams, which allows measuring foams at different drainage states and the control of the rate of foam formation, temperature, and measurement position. The functionality under ESS conditions was tested and neutron test measurement were carried out.

Keywords: foams; small angle neutron scattering; sample environment; instrumentation

1. Introduction

Neutrons play an important role for current fundamental science. Investigation of soft matter in the submicrometer range relies on neutron science due to the possibility of an unique contrast variation based on hydrogen–deuterium exchange. The European Spallation Source (ESS), which is under construction in Lund (Sweden), will be the leading neutron source in terms of flux and brilliance in the future [1,2]. This high flux will lead to decreasing measurement times and, therefore, will reduce the amount of beam time allocated to the individual users. This benefit also comes with a challenge, since the installation of sample environments (SEs) and the time required for preparation of the experiment will become a crucial factor for an efficient use of the beamtime. Consequently, the design of SEs should have a strong emphasis on the reduction of down time during sample and SE changes. The FlexiProb project (funded by the Federal Ministry of Education and Research of Germany, BMBF) is a collaborative effort of three research groups to design and construct three SEs in the field of soft matter research for implementation at the two small angle neutron scattering (SANS) instruments at the ESS, namely LoKI [3] and SKADI [4,5]. The SEs include an in situ SANS/DLS setup [6], a GISANS setup [7], and a SE for SANS experiments on aqueous foams, which is presented in this paper. In principle, the SEs are also compatible with beamlines at other neutron sources. However, the carrier system was designed to specifically fit into the sample areas of the instruments mentioned above. Every SE is assembled on an optical breadboard, which will be mounted on the kinematic mounting system, which is currently developed by the ESS. The whole setup is then transferred to the sample area of SANS instrument with a pallet truck. This approach allows preparing the experimental setup outside the instrument before the actual experiment. In addition, the ESS is planning offline alignment stations at which prealignment can be done within the ESS Universal Sample Coordinate System (USCS) [8]. All of this will allow for an exchange of the whole SE in a single step. This should enable fast and easy changes between different SEs and, therefore, should reduce down time between different experiments and users.

Foams are ubiquitous in everyday life in detergents, cosmetics, and food. In addition, foams are used in industrial processes such as mineral flotation, oil recovery, and fire fighting. Given their large abundance in everyday life and industrial processes, foams are the subject of numerous scientific studies and books [9–15]. However, some fundamental parameters such as foam stability or foamablity (foaming capability) are still difficult to predict [16]. The reason for this lies in the complexity of the foam structure and the vast variety of foam stabilizers ranging from surfactants, ionic and nonionic, to polyelectrolyte/surfactant mixtures, inorganic particles, and proteins. Depending on the system, different parameters seem to govern the overall properties of the resulting macroscopic foam. These parameters include surface elasticity and viscosity [17,18], maximum disjoining pressure in individual foam films [19,20], the formation of aggregates [21,22], and the composition at the interface [23]. The reason for this broad variety of parameters associated with macroscopic foam properties lies not only in the distinct differences between foam stabilizers, but also in the complex structure of the foam itself. All of the above mentioned parameters were studied at single air/water interfaces, bubbles, or foam lamellae, which is a drastic simplification of the complex structure of foams. Although they are challenging, measurements on entire foams can address the structural complexity and dynamics. The dynamics inside of nanoparticle stabilized foams were studied by diffusing wave spectroscopy [24,25]. Here, two dynamic processes were observed: a fast one caused by the nanoparticle diffusion and a slow one reflecting the foam dynamics. The internal structure of foams can be investigated by SANS [26–31]. The main measurable feature here is the thickness of foam lamellae inside of the foam. In this context, specific salt and pH effects [29], the chemical nature of the surfactant [31,32], and different drainage states of the foam [26,29,32] were studied. Furthermore, it is possible to detect objects inside the foam such as micelles [26,27], solid nanoparticles [30], or polymer sufactant complexes [33]. In such experiments, contrast variation is especially powerful, because it allows to selectively mask or unmask objects or structures inside the foam [30]. Since foams are thermodynamically unstable by nature and, with the specific example of aqueous foams, are highly dynamic at a timescale of minutes, SANS experiments were limited to rather stable foams.

Different cell designs for SANS on foams were reported in literature [26,28,29,31]. In all of these cells, foam formation is realized by bubbling gas through a porous plate either made of steel or sintered glass. The first cell of this type was designed by Axelos et al. and consists of a Plexiglas cylinder with a single quartz window for the neutron beam [26]. Taking into account the changing liquid volume fraction along the height of a foam column,

Micheau et al. used a cell with three plane parallel windows at different positions along the foam cylinder [29]. This made it possible to probe the foam at different (gravitational) drainage stages. Following this approach, we designed a foam cell that allows SANS measurements at any desired height along the foam cylinder. Since the foam formation before each SANS experiment takes several minutes, the presented SE includes a sample changer for up to three foam cells, allowing foaming the next sample during the measurement of the previous one. As stated above, this is especially important for highly brilliant neutron sources such as the ESS in order to reduce the downtime between sample changes. The high flux and reduced measurement time at the ESS will also enable measurements with more dynamic and less stable foams and will allow studies with high time resolution, which is especially appealing for highly dynamic foams. In order to use the full potential of the ESS from day one, it is crucial to design and test SEs for various experiments before the start of operation.

In this article, we present the design and construction of the FlexiProb SE for SANS experiments on foams. Detailed technical aspects of the measurement cell itself and the peripheral components such as temperature control, gas flow control, and sample positioning, as well as first benchmarks for measurements with neutrons, are shown.

2. Instrumental Concept and Performance

2.1. General Construction

The main parts of the sample environment are three foam cells, in which the foam will form. Figure 1a,b show an engineering drawing and a photograph of a single foam cell. Each cell consists of a 250 mm long quartz glass cylinder with an inner diameter of 30 mm and a wall thickness of 2 mm. Quartz is almost transparent to neutrons and the circular design avoids the rupture of lamellae at edges. A porous quartz glass plate (pore size 10–16 µm, porosity P16 (ISO 4793)) is fused to the bottom of the cylinder. The whole cylinder is mounted to a gas-inlet socket via an O-ring in a crimp connection. From below, a gas (e.g., air or nitrogen) is pressed through the porous plate, which breaks the continuous gas flow into small bubbles. This results in the formation of foam, when an appropriate foaming solution is poured into the cylinder. Finally, the gas exits the cylinder at the open top, which ensures pressure equilibrium. For temperature control, two thermostating jackets are fitted to each foam cylinder, using titanium screws from the bottom of the gas-inlet socket. One of the main features and advantages of this foam cell is the possibility to measure at any position along the height of the foam cylinder. Therefore, the thermojackets are designed in a way to leave a slit-shaped gap between them. The width of this gap governs the maximum scattering angle accessible. Figure 1c shows a sketch in top view for the estimation of the maximum scattering angle at the geometrically least favorable position. For a scattering event occurring at the edge of a 10 mm wide primary neutron beam (highlighted in orange) at the side facing the primary beam, the scattering angle limit $2\theta_{lim}$ is 13.8°. Assuming a neutron wavelength of $\lambda = 5$ Å, the maximum scattering vector q_{max} is calculated by

$$q_{max} = \frac{4\pi}{\lambda} \sin(\theta_{lim}) \approx 0.3 \, \text{Å}^{-1} \tag{1}$$

This corresponds to a minimum measurable size of 2 nm in real space, which is sufficient to resolve structures such as foam films or incorporated objects such as micelles or nanoparticles. It is worth noting that the SANS instruments LoKI and SKADI at the ESS will be operated with a predicted neutron wavelength band of 2 Å to 22 Å and 3 Å to 21 Å, respectively [2]. We expect an accessible q range of 0.005–0.6 Å^{-1} for our foam cell at these ESS instruments. The lower limit is governed by the instrument's design and the upper limit by the cell geometry. This will lead to a range of measurable sizes of ca. 1–125 nm. The gas-inlet socket and thermostating cylinder are made of $AlMg_{4.5}Mn_{0.7}$, a special aluminum alloy that is easy to process and not activated by neutrons. The hose connectors are made of $AlMg_{4.5}$ and, therefore, are also not activated by neutrons. All three foam cells are placed in a folded $AlMg_{4.5}$ socket.

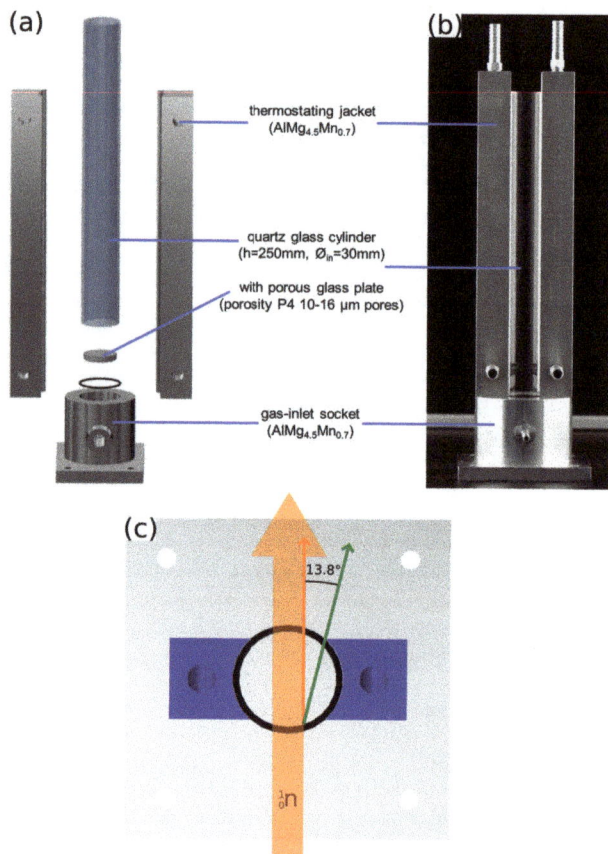

Figure 1. Individual foam cell. (**a**) Explosion-view drawing and (**b**) photograph in front view. (**c**) Schematic top view. The individual foam cells consist of a quartz glass cylinder, thermojackets, and a gas-inlet socket. The neutron beam (highlighted in orange) passes the cell perpendicular to the thermojackets in variable height. For the estimation of the minimal range of scattering angles accessible, a scattering event at the (geometrically) least favorable position is considered (green arrow).

The side facing the primary neutron beam is protected by an aluminum plate covered with 1 mm B_4C including rectangular slits at the respective positions of the foam cells. Since the cells are open to the top, an aluminum cover was placed above the cells to avoid dust incorporation. A backview of the mounted cells is shown in Figure 2a. The setup is placed on a translation stage (travel range 508 mm), which allows remote controlled sample changes between the different foam cells as shown in Figure 2b,c. The entire setup is mounted on a breadboard (900 × 1200 mm^2, Newport Spectra-Physics GmbH, Darmstadt, Germany). Later on at the ESS, the entire setup will be mounted on a lifting table, which is also part of the unified carrier system used by all FlexiProb SEs. This will ensure fast setup changes by removing the entire setup including the board. The lifting table installed at LoKI and SKADI will have a vertical displacement range of >300 mm with a 0.1 mm positional accuracy, which is sufficient for the foam cell SE.

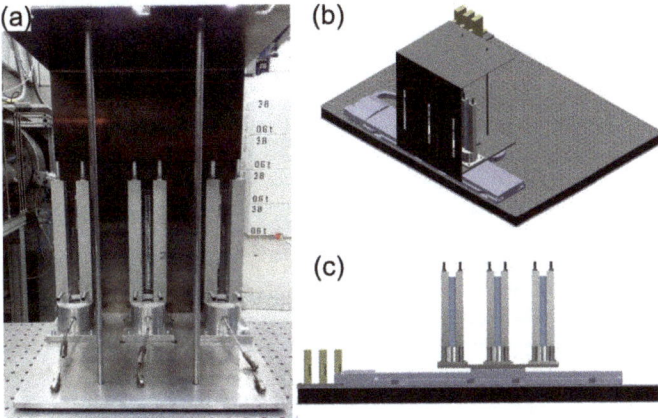

Figure 2. Overview of the sample environment (SE). (**a**) Backview of three foam cells on folded aluminum construction. (**b**) Technical drawing of the complete SE mounted on the breadboard. (**c**) Front view of the SE (technical drawing) without shielding.

2.2. Gas Flow Control

The flow of the foaming gas is controlled by a FG-201CV mass flow controller (Bronkhorst, AK Ruurlo, The Netherlands) operating at an upstream pressure of 3 bar. The gas flow rate can be adjusted between 0–30 mL min^{-1}. The three foam cells have separate gas circuits with individual mass flow controllers. This opens the possibility to prepare foams while another is measured or to define different foam states with a fast shift between different columns. The devices are connected via RS232 connections to a serial device server (NPort 5450, Moxa, Taipei, Taiwan). Here, the RS232 signal is converted into an Ethernet signal, which is transmitted via LAN to the instrument control. Communication is realized on a dynamic data exchange (DDE) server implemented in the supplier's FlowDDE software (Version 4.81). With this connection established, the devices are controlled using the DDE client program FlowView (Version 1.23) also distributed by the supplier.

Figure 3 shows the results of a foaming experiment in which 12 mL of tetradecyltrimethylammonium bromide (C_{14}TAB) ($c = 3.5$ mM) surfactant solution were foamed with synthetic air at a flow rate of $\dot{V} = 30$ mL min^{-1}. Pictures of the foam column were taken in a time interval of 80 s and the respective foam heights extracted by image analysis. Neglecting foam decay, a theoretical foam height $h_{th}(t)$ was calculated using Equation (2) for comparison, which is based on the volumetric gas flow rate and the cross-sectional area of the foam cell.

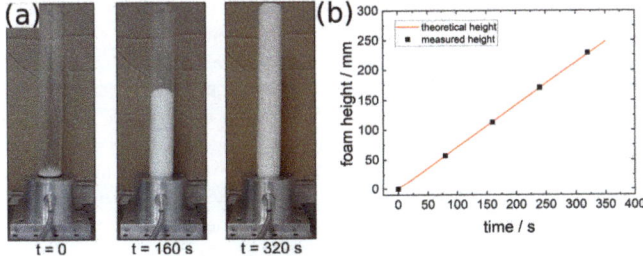

Figure 3. Test of the foaming procedure and tightness of the gas flow system. (**a**) Three pictures of a foaming C_{14}TAB ($c = 3.5$ mM) solution at different times. (**b**) Foam height as a function of time (black squares) and the theoretical height (red line) according to Equation (2).

$$h_{th}(t) = \frac{\dot{V} \cdot t}{A} \qquad (2)$$

Here, $\dot{V}$ is the volumetric gas flow rate, t is the foaming time, and A is the cross-sectional area of the foaming cylinder.

The measured values are in good agreement with the theoretical foam height as shown in Figure 3b. This shows that the gas flow system works and the foam is formed as intended.

2.3. Temperature Control

Temperature change of the foams is achieved by two $AlMg_{4.5}Mn_{0.7}$ thermostating jackets at both sides of each foam cylinder. The thermostating jackets of all three sample cells are connected via appropriate distributors to a Julabo FP50-HL circulating thermostat (Julabo GmbH, Seelbach, Germany) with a temperature range of $-50\,°C$ to $200\,°C$ when the appropriate thermofluid is used. For studying aqueous foams, however, an achievable temperature range of $10\,°C$ to $80\,°C$ is typically sufficient, allowing the use of water as thermofluid. To validate the performance of the heat input by the thermostating jackets, tests with a foam stabilized by C_{14}TAB at its critical micelle concentration of $c = 3.5$ mM were performed [34]. Therefore, 12 mL of the surfactant solution were foamed at ambient temperature with a synthetic air flow of $15\,mL\,min^{-1}$. After the foam reached the top of the quartz cylinder, the gas flow was stopped and the setpoint of the water bath thermostat was adjusted to $50\,°C$. The evolution of the temperature was monitored by three Pt-100 temperature sensors (model PT-102-3S-QT, Lake Shore Cryotronics, Inc., Westerville, OH, USA) at different positions at a height of 11.5 cm inside of the foam cylinder. The positions were chosen in a way to reflect the asymmetric shape of the thermostating jacket around the foam cylinder as indicated by A, B, and C in Figure 4. One sensor was put in the center of the cylinder (C), while the remaining two sensors were placed on the rim of the foam column, one in close proximity to the thermostating jacket (A) and one right behind the neutron window slit (B). The evolution of the temperature and a schematic sketch of the different positions are shown in Figure 4a.

Following an initial increase, the temperature reaches a plateau at all three positions after around 20 min. The final temperatures were $48\,°C$ at position A (close to thermostating jacket), $42\,°C$ at position B (behind neutron window slit), and $45\,°C$ at position C (center). This temperature gradient is explained by the asymmetric shape of the thermostating jacket around the foam cell and the low heat conductivity of foams. As explained in Section 2.1, this asymmetric shape was chosen to allow SANS measurements at any height along the foam cylinder, accepting the drawback of a potential temperature gradient. The temperature jumps observed at every position are most likely due to air bubbles passing by the sensors, changing the heat conductivity next to them.

In order to reduce this temperature gradient, the experiment was repeated with the foam cylinder wrapped in aluminum foil (see Figure 4b). Again, the temperature reaches a plateau after around 20 min with final temperatures of $49\,°C$ at position A, $47\,°C$ at position B, and $48\,°C$ at position C. Wrapping the cylinder in aluminum foil decreases the temperature gradient inside the foam because of the improved thermal contact between the quartz cylinder and the thermostating jackets. In addition, aluminum is almost transparent to neutrons and is also often used as a material for neutron windows. A drawback of this approach is that it is no longer possible to observe the foam during a SANS experiment with a camera. Detecting the formation of holes at the measuring position is sometimes beneficial, especially when dealing with rather unstable foams where holes may form randomly. Depending on the requirements regarding the accuracy of temperature control, one of the two methods described can be used.

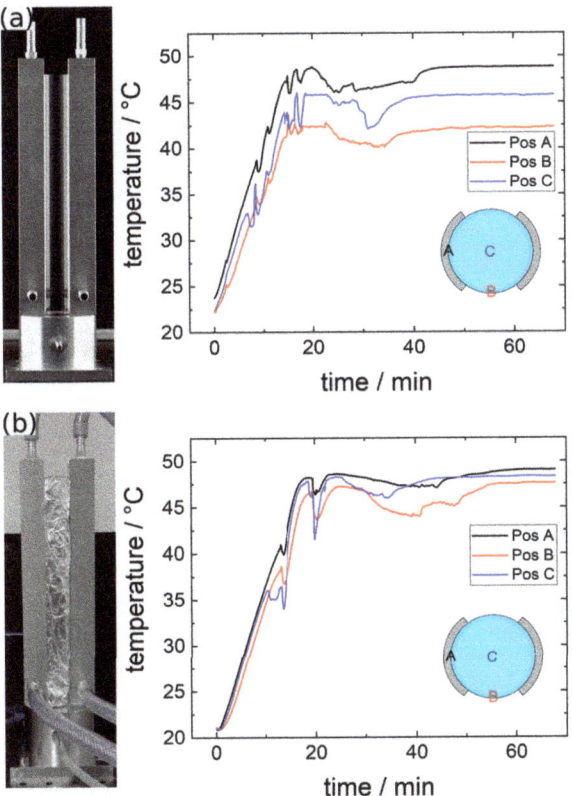

Figure 4. Evolution of the temperature of a tetradecyltrimethylammonium bromide (C_{14}TAB, $c = 3.5$ mM) foam in the foam cylinder without (**a**) and with (**b**) aluminum foil cover. The temperature was monitored using three Pt-100 temperature sensors at positions depicted by the inset. The setpoint of the water bath thermostat was adjusted to 50 °C at the beginning of the experiment and was reached after ca. 8 min.

2.4. Sample Positioning

Horizontal sample alignment is achieved with a linear translation stage (LS-180, Physik Instrumente GmbH & Co. KG, Karlsruhe, Germany) with a maximum load of 100 kg, an operating displacement of 508 mm at a maximum speed of 150 mm s^{-1}, and a bidirectional repetition accuracy of ±0.1 µm. The linear stage is driven by a two-phase bipolar half-coil stepper motor (model PK-258-02B, Oriental Motor, Tokyo, Japan). The position is monitored with a linear optical encoder with RS-422 quadrature signal transmission (LIA-20, Numerik Jena, Jena, Germany). The entire setup will be placed on an optical breadboard based on a lifting table, which ensures the vertical sample alignment. The control unit was built according to ESS specifications ensuring compatibility with ESS control standards. Figure 5 shows the corresponding circuit diagram for the custom built crate. The translation stage is labeled with "AXIS 1". For testing purposes, a two-axis goniometer was also integrated (motor top, motor bottom, encoder top, and encoder bottom). However, in the framework of the foam cell SE, this goniometer is not used.

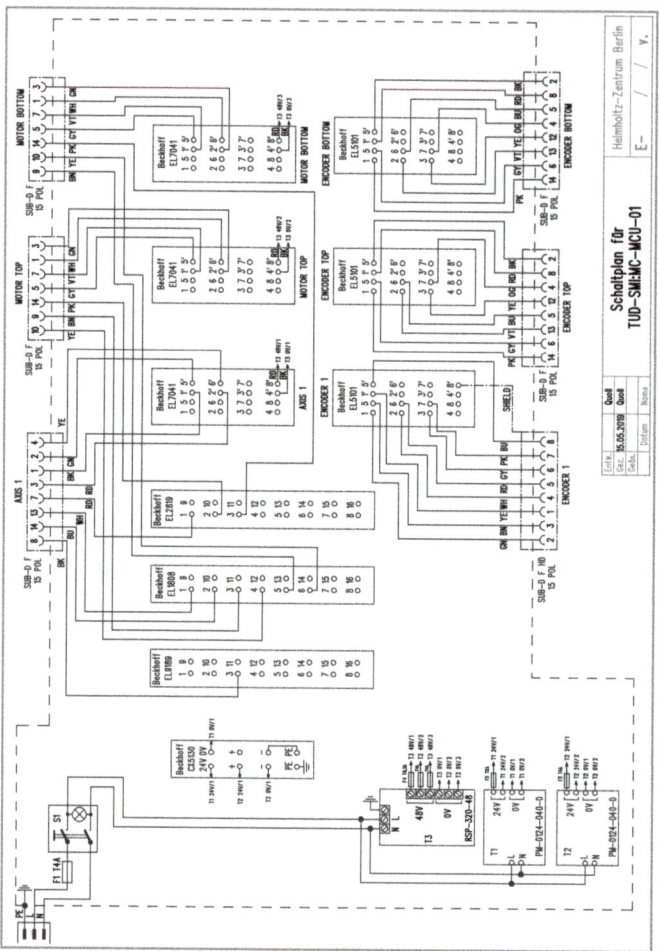

Figure 5. Circiut diagram of the motion control crate. Beside the translation stage, a two-axis goniometer is also controlled by this crate.

The custom built 19-inch crate is equipped with 24 V (BLOCK Transformatoren-Elektronik GmbH, Verden, Germany) and 48 V (Mean Well Enterprises, New Taipei City, Taiwan) power supplies. The latter one provides power for the motors while the first one provides the power for the control unit. The control unit is based on an Ethernet fieldbus system (EtherCAT, Beckhoff Automation GmbH & Co. KG, Verl, Germany). An embedded PC (CX5130) is connected to a potential distribution terminal (EL9189) and a stepper motor terminal (EL7041), which ensure the motor movement. A digital input terminal (EL1808) and a digital output terminal (EL2819) are connected, sending and reading the motor positions. Additionally, an incremental encoder interface terminal (EL5101) is added to read the encoder signal out.

The embedded PC communicates with the hardware of the linear stage via the Twin-CAT 3 software. This signal is forwarded to the Experimental Physics and Industrial Control System (EPICS) [35], which will be the unified control software for beamline devices at the ESS and is able to control the devices and monitor their state. The EPICS is used by the Networked Instrument Control System (NICOS) [36], which offers a graphical interface for users and is the outermost layer of the instrument control structure at the ESS. Further details regarding instrument control and data streaming are described elsewhere [37].

From 2015 until 2019, the ESS operated a dedicated testbeamline at the Helmholtz-Zentrum Berlin, also known as V20 [38,39]. Here, a fully workable environment, mimicking a future ESS instrument, was built up for testing and development. The linear stage was successfully tested at this instrument, proving the compatibility with the ESS standards in terms of control, interaction with other devices, and data logging.

The integration of the peripheral SE components into NICOS also allows programmable measurements with automated changes between the three foam cells and the measurement height along each cell, varying temperature, and gas flows.

3. Neutron Test Measurements

Test measurements with a single foam cell were carried out at the KWS-1 small angle scattering diffractometer at the Heinz Maier-Leibnitz-Zentrum (MLZ, Garching, Germany) [40,41]. Figure 6a shows the foam cell at the beamline. All measurements were performed at a wavelength of 4.92 Å with a 10% wavelength resolution (FWHM), a squared neutron beam of 10×10 mm, and a data acquisition time of 5 min. Scattering patterns were recorded with a ^{6}Li-scintillation detector with photomultiplier tubes and a spatial resolution of 5.3×5.3 mm^2. All measurements were background corrected due to dark current. The foam measurements were also corrected for the scattering by the empty cell. The sample–detector distance was determined using an optical theodolite. Figure 6b shows the 2D detector image of the empty foam cell recorded at a sample–detector distance of 7.615 m. As expected, the empty quartz cylinder provides a low background with no significant secondary scattering.

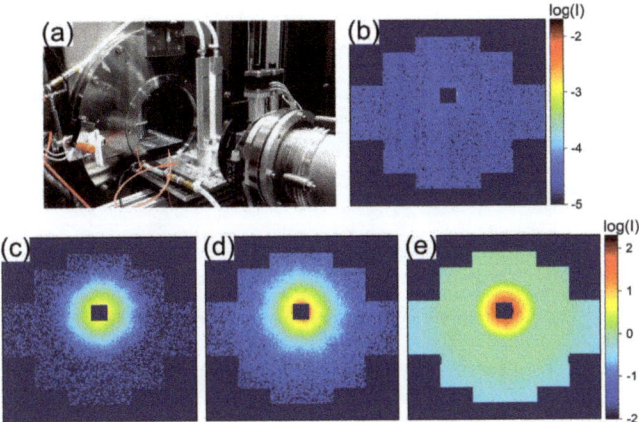

Figure 6. (a) Single foam cell installed at the KWS-1 beamline at the MLZ. (b) 2D SANS data of an empty quartz foam cylinder. (c–e) 2D SANS data of a steady-state foam produced from a 25 g L^{-1} SDS solution at 16 cm (c), 9.5 cm (d), and 2 cm (e) foam height above the foaming solution. All experiments were carried out at a sample–detector distance of 7.615 m. Data acquisition time was 5 min.

Test measurements were conducted with a foam stabilized by 25 g L^{-1} (86.7 mM) sodium dodecyl sulfate (SDS), which is well above the critical micelle concentration of 8 mM [42], at different foam heights. This system was chosen because of the high foam stability and the fact that the first study ever performed on SANS on foams also used a SDS foam at this concentration, making it a reference system [26]. 12 mL of the surfactant solution were initially foamed with a nitrogen gas flow rate of 10 mL min^{-1}. After the foam level reached a height of around 18 cm, the gas flow was reduced to 1 mL min^{-1}. At this flow rate the foam height does not change anymore, meaning that the foam formation at the bottom and the foam decay at the top of the column are balanced. This results in a steady-state foam, in which the foam height corresponds to the age of the foam (i.e., the time passed after its formation at the bottom of the cylinder) and therefore its drainage state.

Figure 6c–e shows the corresponding 2D detector images recorded at 16 cm, 9.5 cm and 2 cm above the foaming solution. All images exhibit an isotropic scattering signal around the primary beam. This reveals that the neutron path length through the cell is long enough to average over all orientations of the foam structure, namely the liquid films, within the measuring window. The scattered intensity decreases with increasing measurement height. This is explained by the decreasing liquid volume fraction of the foam with increasing foam age (or height) and proves that different states of the foam can be accessed with a steady-state foam in one experiment.

Axelos et al. studied the same system using a similar foam cell with a single quartz window for the neutron beam in the lower third of the cylinder and the reported values are shown for comparison (closed squares in Figure 7) [26]. They performed two types of measurements. The first type of measurement was a continuous foaming experiment, in which the SANS data were recorded while nitrogen was continuously bubbled through the foaming solution. This resulted in a wet foam at the measuring position. The second type of measurement was a drainage experiment, in which the gas flow was stopped after foam formation. After some time, this led to a dry foam at the measuring position at the bottom of the foam column. These two types of foams are related to different foam heights in a steady-state foam. Here, a wet and fresh foam is observed at the bottom while a dry and aged foam is observed at the top of the steady-state foam.

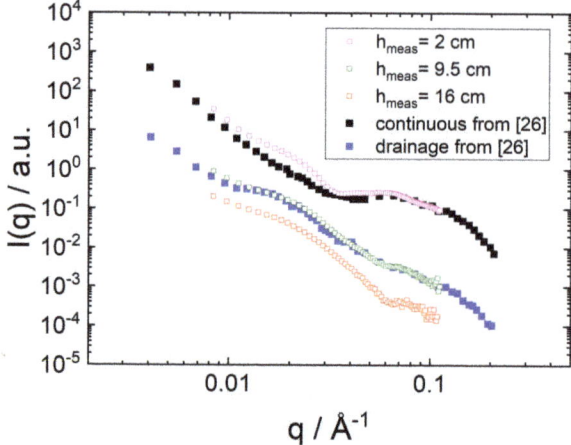

Figure 7. Radial averaged scattering data from a 25 g L^{-1} SDS foam (86.7 mM) at three different foam heights in a steady-state foam (open symbols) and comparison with reference data of a wet continuous foam and a foam after drainage [26].

The measured radial averaged scattering curves are plotted in Figure 7 (open squares). The data were shifted in intensity for better comparison. Normalization of SANS data of foams is still challenging and was not an aim of this experiment. The determination of the foam's liquid volume fraction based on the neutron transmission is prone to errors, since most foams show a high transmission close to 1 [26,29]. In general, the trend of measured scattering curves are in good agreement with the literature data, showing that there are no intrinsic artefacts caused by the foam cell. Data recorded at a height of 2 cm are similar to the one of a wet foam reported in literature and the data recorded at a height of 9.5 cm are in good agreement with the one of a drained foam. The data measured at a height of 16 cm correspond to a foam in an even more drained state. The shoulder at $q \approx 0.08$ Å^{-1} is attributed to surfactant micelles present in the liquid foam films. Consequently, the intensity of this shoulder decreases with increasing foam height, as the liquid drains out of the foam films over time. The shoulder at $q \approx 0.02$ Å^{-1} occurs due to scattering at the liquid foam films and is related to their thickness [26]. The fact that this shoulder was not observed by Axelos et al. in their wet continuous foam could be explained by an even lower measuring position or higher gas flow rate in the steady-state. Both factors would results in a wetter foam, leading to a less ordered foam structure with a higher polydispersity regarding the film thickness and, therefore, in a smearing out of the corresponding scattering feature. Since it was possible to reproduce two different experiments reported in the literature by scanning along the foam height of a steady-state foam, the presented approach is valid for accessing the foam in different drainage stages with varying liquid volume fractions in one experiment.

4. Conclusions

We successfully designed, constructed, and tested a sample environment (SE) for SANS measurements on liquid foams. The SE allows the control of the gas flow rate used for foam formation and offers the possibility to control the temperature. The complete foam cylinder is made from neutron-transparent quartz glass, which enables SANS measurements at any position along the foam height and the possibility to study different drainage stages in a single experiment. Additionally, a sample changer on the basis of a linear stage was constructed and completely integrated into the foreseen ESS control structure at the ESS test beamline V20 at the HZB. Finally, SANS test measurements were successfully performed at the KWS-1 beamline at the MLZ with a model foam already reported in literature.

The presented SE is well-suited for studying liquid foams with SANS and should meet the special requirements of neutron sources with high brilliance such as the ESS, where an efficient use of the allocated beamtime becomes more important and the time required for sample changes should be reduced to an absolute minimum. This is accounted for by the sample changer for up to three foam cells and the possibility to access different states of the same foam in a single measurement cell. In return, a neutron source such as the ESS will allow new and/or more detailed SANS experiments using foams. The high neutron flux at the ESS will reduce the data acquisition time for a similar experiment as presented above at least by one order of magnitude to below 1 min. This will enable experiments with less stable foams or studies investigating structural changes in foams caused by external stimuli.

Author Contributions: Conceptualization, T.H., P.M.-B., and R.v.K.; planning and discussion of setup components, M.K., O.L., T.W., L.P.K., A.J.S., L.W., S.J., H.F., H.S., and A.H.; construction and test of setup components, M.K. and O.L.; software integration, M.K., O.L., T.B., and P.B.; SANS measurements, M.K., O.L., and H.F.; writing—original draft preparation, M.K.; writing—review and editing, O.L., T.W., L.P.K., A.J.S., L.W., S.J., H.F., T.B., P.B., H.S., A.H., T.H., P.M.-B., and R.v.K.; visualization, M.K.; supervision, O.L. and R.v.K.; project administration, T.H.; funding acquisition, T.H., P.M.-B., and R.v.K. All authors have read and agreed to the published version of the manuscript.

Funding: This research was funded by the German Federal Ministry for Education and Research (BMBF) within the project "FlexiProb" sample environment grant number 05K2016.

Institutional Review Board Statement: Not applicable.

Informed Consent Statement: Not applicable.

Data Availability Statement: Derived data supporting the findings of this study are available from the corresponding author upon reasonable request.

Acknowledgments: We thank Judith Houston (ESS) for valuable discussions and Dirk Oppermann (TU Darmstadt) for help with constructing the foam cell. Furthermore, We thank Ibrahim El-Idrisi (TU Darmstadt) and Konstantin Quoll (HZB) for wiring and testing the motion control crate. We would also like to acknowledge Robin Woracek (ESS) and Peter Kadletz (ESS) for the help at the ESS test beamline.

Conflicts of Interest: The authors declare no conflict of interest.

References

1. Garoby, R.; Vergara, A.; Danared, H.; Alonso, I.; Bargallo, E.; Cheymol, B.; Darve, C.; Eshraqi, M.; Hassanzadegan, H.; Jansson, A.; et al. The European Spallation Source Design. *Phys. Scr.* **2017**, *93*, 014001. [CrossRef]
2. Andersen, K.; Argyriou, D.; Jackson, A.; Houston, J.; Henry, P.; Deen, P.; Toft-Petersen, R.; Beran, P.; Strobl, M.; Arnold, T.; et al. The instrument suite of the European Spallation Source. *Nucl. Instrum. Methods Phys. Res. Sect. A* **2020**, *957*, 163402. [CrossRef]
3. Jackson, A.J.; Kanaki, K. ESS Construction Proposal LoKI—A Broad-Band SANS Instrument. 2013. Available online: https://europeanspallationsource.se/sites/default/files/files/document/2017-09/loki_proposal_stc_sept2013.pdf (accessed on 12 April 2021).
4. Jaksch, S.; Martin-Rodriguez, D.; Ostermann, A.; Jestin, J.; Duarte Pinto, S.; Bouwman, W.G.; Uher, J.; Engels, R.; Frielinghaus, H. Concept for a time-of-flight Small Angle Neutron Scattering instrument at the European Spallation Source. *Nucl. Instrum. Methods Phys. Res. Sect. A* **2014**, *762*, 22–30. [CrossRef]
5. Jaksch, S.; Chennevière, A.; Désert, S.; Kozielewski, T.; Feilbach, H.; Lavie, P.; Hanslik, R.; Gussen, A.; Butterweck, S.; Engels, R.; et al. Technical Specification of the Small-Angle Neutron Scattering Instrument SKADI at the European Spallation Source. *Appl. Sci.* **2021**, *11*, 3620. [CrossRef]
6. Schmid, A.J.; Wiehemeier, L.; Jaksch, S.; Schneider, H.; Hiess, A.; Bögershausen, T.; Widmann, T.; Reitenbach, J.; Kreuzer, L.P.; Löhmann, O.; et al. Flexible Sample Environments for the Investigation of Soft Matter at the European Spallation Source: Part I—The In Situ SANS/DLS Setup. *Appl. Sci.* **2021**, *11*, 4089. [CrossRef]
7. Widmann, T.; Kreuzer, L.P.; Kühnhammer, M.; Schmid, A.J.; Wiehemeier, L.; Jaksch, S.; Frielinghaus, H.; Löhmann, O.; Schneider, H.; Hiess, A.; et al. Flexible Sample Environment for the Investigation of Soft Matter at the European Spallation Source: Part II—The GISANS Setup. *Appl. Sci.* **2021**, *11*, 4036. [CrossRef]
8. Guthrie, M.; Perez, B. ESS Sample Environment Mechanical Interfaces for Instruments. 2015. Available online: https://indico.esss.lu.se/event/915/attachments/7041/10120/ESS-0038078_Mech_Int_20170919_V2_for_distribution.pdf (accessed on 18 May 2021).
9. Weaire, D.; Hutzler, S. *The Physics of Foams*; Oxford University Press: Oxford, UK; New York, NY, USA, 1999.
10. Exerowa, D.R.; Gochev, G.; Platikanov, D.; Liggieri, L.; Miller, R. *Foam Films and Foams—Fundamentals and Applications*; CRC Press: Boca Raton, FL, USA, 2019.
11. Langevin, D. Aqueous foams: A field of investigation at the Frontier between chemistry and physics. *ChemPhysChem* **2008**, *9*, 510–522. [CrossRef]
12. Fameau, A.L.; Carl, A.; Saint-Jalmes, A.; von Klitzing, R. Responsive aqueous foams. *ChemPhysChem* **2015**, *16*, 66–75. [CrossRef]
13. Hill, C.; Eastoe, J. Foams: From nature to industry. *Adv. Colloid Interface Sci.* **2017**, *247*, 496–513. [CrossRef]
14. Horozov, T. Foams and foam films stabilised by solid particles. *Curr. Opin. Colloid Interface Sci.* **2008**, *13*, 134–140. [CrossRef]
15. Drenckhan, W.; Saint-Jalmes, A. The science of foaming. *Adv. Colloid Interface Sci.* **2015**, *222*, 228–259. [CrossRef]
16. Braun, L.; Kühnhammer, M.; von Klitzing, R. Stability of aqueous foam films and foams containing polymers: Discrepancies between different length scales. *Curr. Opin. Colloid Interface Sci.* **2020**, *50*, 101379. [CrossRef]
17. Hofmann, M.J.; Motschmann, H. A parameter predicting the foam stability of mixtures of aqueous ionic amphiphile solutions with indifferent electrolyte. *Colloids Surf. A Physicochem. Eng. Asp.* **2017**, *529*, 1024–1029. [CrossRef]
18. Stubenrauch, C.; Miller, R. Stability of foam films and surface rheology: An oscillating bubble study at low frequencies. *J. Phys. Chem. B* **2004**, *108*, 6412–6421. [CrossRef]
19. Stubenrauch, C.; Klitzing, R.V. Forces and structure in thin liquid soap films Disjoining pressure in thin liquid foam and emulsion films—New concepts and perspectives. *J. Phys. Condens. Matter* **2003**, *15*, R1197–R1232. [CrossRef]
20. Fauser, H.; von Klitzing, R. Effect of polyelectrolytes on (de)stability of liquid foam films. *Soft Matter* **2014**, *10*, 6903–6916. [CrossRef]
21. Carl, A.; Bannuscher, A.; Von Klitzing, R. Particle stabilized aqueous foams at different length scales: Synergy between silica particles and alkylamines. *Langmuir* **2015**, *31*, 1615–1622. [CrossRef] [PubMed]
22. Braunschweig, B.; Schulze-Zachau, F.; Nagel, E.; Engelhardt, K.; Stoyanov, S.; Gochev, G.; Khristov, K.; Mileva, E.; Exerowa, D.; Miller, R.; et al. Specific effects of Ca^{2+} ions and molecular structure of β-lactoglobulin interfacial layers that drive macroscopic foam stability. *Soft Matter* **2016**, *12*, 5995–6004. [CrossRef] [PubMed]
23. Uhlig, M.; Löhmann, O.; Vargas Ruiz, S.; Varga, I.; von Klitzing, R.; Campbell, R.A. New structural approach to rationalize the foam film stability of oppositely charged polyelectrolyte/surfactant mixtures. *Chem. Commun.* **2020**, *56*, 952–955. [CrossRef]
24. Crassous, J.; Saint-Jalmes, A. Probing the dynamics of particles in an aging dispersion using diffusing wave spectroscopy. *Soft Matter* **2012**, *8*, 7683–7689. [CrossRef]
25. Carl, A.; Witte, F.; von Klitzing, R. A look inside particle stabilized foams—Particle structure and dynamics. *J. Phys. D Appl. Phys.* **2015**, *48*, 434003. [CrossRef]
26. Axelos, M.A.V.; Boué, F. Foams as Viewed by Small-Angle Neutron Scattering. *Langmuir* **2003**, *19*, 6598–6604. [CrossRef]
27. Fameau, A.L.; Saint-Jalmes, A.; Cousin, F.; Houinsou Houssou, B.; Novales, B.; Navailles, L.; Nallet, F.; Gaillard, C.; Boué, F.; Douliez, J.P. Smart foams: Switching reversibly between ultrastable and unstable foams. *Angew. Chem. Int. Ed.* **2011**, *50*, 8264–8269. [CrossRef] [PubMed]

28. Hurcom, J.; Paul, A.; Heenan, R.K.; Davies, A.; Woodman, N.; Schweins, R.; Griffiths, P.C. The interfacial structure of polymeric surfactant stabilised air-in-water foams. *Soft Matter* **2014**, *10*, 3003–3008. [CrossRef]
29. Micheau, C.; Bauduin, P.; Diat, O.; Faure, S. Specific Salt and pH Effects on Foam Film of a pH Sensitive Surfactant. *Langmuir* **2013**, *29*, 8472–8481. [CrossRef] [PubMed]
30. Mikhailovskaya, A.; Zhang, L.; Cousin, F.; Boué, F.; Yazhgur, P.; Muller, F.; Gay, C.; Salonen, A. Probing foam with neutrons. *Adv. Colloid Interface Sci.* **2017**, *247*, 444–453. [CrossRef]
31. Yada, S.; Shimosegawa, H.; Fujita, H.; Yamada, M.; Matsue, Y.; Yoshimura, T. Microstructural Characterization of Foam Formed by Hydroxy Group–Containing Amino Acid Surfactant Using Small-Angle Neutron Scattering. *Langmuir* **2020**. [CrossRef] [PubMed]
32. Mansour, O.T.; Cattoz, B.; Beaube, M.; Montagnon, M.; Heenan, R.K.; Schweins, R.; Appavou, M.S.; Griffiths, P.C. Assembly of small molecule surfactants at highly dynamic air-water interfaces. *Soft Matter* **2017**, *13*, 8807–8815. [CrossRef]
33. Mansour, O.T.; Cattoz, B.; Beaube, M.; Heenan, R.K.; Schweins, R.; Hurcom, J.; Griffiths, P.C. Segregation versus interdigitation in highly dynamic polymer/surfactant layers. *Polymers* **2019**, *11*, 109. [CrossRef]
34. Domínguez, A.; Fernández, A.; Gonzalez, N.; Iglesias, E.; Montenegro, L. Determination of critical micelle concentration of some surfactants by three techniques. *J. Chem. Educ.* **1997**, *74*, 1227–1231. [CrossRef]
35. EPICS. Available online: https://epics-controls.org/ (accessed on 12 April 2021).
36. NICOS. Available online: https://nicos-controls.org/ (accessed on 12 April 2021).
37. Löhmann, O.; Silvi, L.; Kadletz, P.M.; Vaytet, N.; Arnold, O.; Jones, M.D.; Nilsson, J.; Hart, M.; Richter, T.; von Klitzing, R.; et al. Wavelength frame multiplication for reflectometry at long-pulse neutron sources. *Rev. Sci. Instrum.* **2020**, *91*, 125111. [CrossRef]
38. Strobl, M.; Bulat, M.; Habicht, K. The wavelength frame multiplication chopper system for the ESS test beamline at the BER II reactor—A concept study of a fundamental ESS instrument principle. *Nucl. Instrum. Methods Phys. Res. Sect. A* **2013**, *705*, 74–84. [CrossRef]
39. Woracek, R.; Hofmann, T.; Bulat, M.; Sales, M.; Habicht, K.; Andersen, K.; Strobl, M. The test beamline of the European Spallation Source—Instrumentation development and wavelength frame multiplication. *Nucl. Instrum. Methods Phys. Res. Sect. A* **2016**, *839*, 102–116. [CrossRef]
40. Frielinghaus, H.; Feoktystov, A.; Berts, I.; Mangiapia, G. KWS-1: Small-angle scattering diffractometer. *J. Large-Scale Res. Facil. JLSRF* **2015**, *1*, 26–29. [CrossRef]
41. Feoktystov, A.V.; Frielinghaus, H.; Di, Z.; Jaksch, S.; Pipich, V.; Appavou, M.S.; Babcock, E.; Hanslik, R.; Engels, R.; Kemmerling, G.; et al. KWS-1 high-resolution small-angle neutron scattering instrument at JCNS: Current state. *J. Appl. Crystallogr.* **2015**, *48*, 61–70. [CrossRef]
42. Danov, K.D.; Kralchevsky, P.A.; Ananthapadmanabhan, K.P. Micelle-monomer equilibria in solutions of ionic surfactants and in ionic-nonionic mixtures: A generalized phase separation model. *Adv. Colloid Interface Sci.* **2014**, *206*, 17–45. [CrossRef] [PubMed]

Review

Soft Matter Sample Environments for Time-Resolved Small Angle Neutron Scattering Experiments: A Review

Volker S. Urban [1,*], William T. Heller [1], John Katsaras [1] and Wim Bras [2,*]

[1] Neutron Scattering Division, Oak Ridge National Laboratory, One Bethel Valley Road, Oak Ridge, TN 37831, USA; hellerwt@ornl.gov (W.T.H.); katsarasj@ornl.gov (J.K.)
[2] Chemical Sciences Division, Oak Ridge National Laboratory, One Bethel Valley Road, Oak Ridge, TN 37831, USA
* Correspondence: urbanvs@ornl.gov (V.S.U.); brasw@ornl.gov (W.B.)

Featured Application: This manuscript has been authored by UT-Battelle, LLC, under contract DE-AC05-00OR22725 with the US Department of Energy (DOE). The US government retains and the publisher, by accepting the article for publication, acknowledges that the US government retains a nonexclusive, paid-up, irrevocable, worldwide license to publish or reproduce the published form of this manuscript, or allow others to do so, for US government purposes.

Abstract: With the promise of new, more powerful neutron sources in the future, the possibilities for time-resolved neutron scattering experiments will improve and are bound to gain in interest. While there is already a large body of work on the accurate control of temperature, pressure, and magnetic fields for static experiments, this field is less well developed for time-resolved experiments on soft condensed matter and biomaterials. We present here an overview of different sample environments and technique combinations that have been developed so far and which might inspire further developments so that one can take full advantage of both the existing facilities as well as the possibilities that future high intensity neutron sources will offer.

Keywords: soft matter; neutron scattering; time-resolved; sample environments

1. Introduction

The use of neutron scattering in soft materials research has a long tradition. Neutrons scatter from atomic nuclei and neutron scattering lengths depend unsystematically on the atomic number, and in the case of hydrogen-rich soft materials, the scattering lengths of protium and its isotope, deuterium, are very different. This ability to vary neutron contrast in hydrogen-rich materials with H/D isotopic labels has made neutron scattering an important tool in elucidating some of the basic concepts of polymer dynamics, phase behavior, and molecular conformation, both in solution and in the bulk.

There are a limited number of neutron sources and user facilities, which limits the number of experiments that can be performed relative to those possible using X-ray scattering techniques. Yet neutrons remain vital for materials research, owing to the unique information provided by their unique atomic and isotopic sensitivity, magnetic moment, and highly penetrating nature. An excellent opportunity exists to develop time-resolved experiments and the sample environments needed for these experiments. Related developments have taken place in X-ray synchrotron radiation facilities, in part due to the abundance of beamlines [1–3]. Despite the limitations due to the limited number of neutron scattering beamlines, the scope for developments of time-resolved experiments and the required sample environments is much wider than often is perceived.

The impact that neutron scattering can have on soft matter research can be increased when the capabilities of neutrons as applied to static structure determinations can be extended to the time-resolved domain. Apart from the data acquisition systems needed to enable such experiments, it is also required to provide sample environments capable of perturbing samples in a controlled and homogeneous manner.

For soft matter studies, it is not only important to control sample parameters, such as temperature, pressure, pH etc. accurately, but it is also necessary to control the parameter history. For instance, the different thermal histories for a given material can lead to very different morphologies. This is particularly relevant when using complicated sample environments that try to mimic the conditions in which a material is exposed to processing conditions. Here, the combination of temperature, volume, pressure, shear force etc. has to be accurately controlled in order to be able to obtain meaningful insights into the material behavior during the process.

Neutron scattering is rarely the sole tool used to obtain an understanding of samples in out-of-equilibrium conditions and is often used in combination with data sets obtained from other experimental techniques where each technique addresses different length scales or thermodynamic aspects of the problem. If the sample environment is complicated enough and if the time-resolution is so fast that it becomes difficult to combine the data sets from individual experiments at a later stage, it makes sense to combine such experiments in a single multimodal experiment. This adds to the experimental complexity, but the benefit of obtaining complementary data, knowing that the sample is in the same physical state, outweighs the extra required efforts in developing and implementing such sample environments, and represents an excellent growth opportunity for novel neutron scattering instrumentation. Importantly, the availability of multi-modal experiments can increase the productivity and scientific impact of the large-scale neutron scattering facilities in a more cost-effective way than projects aimed solely at increasing neutron flux.

This article provides an overview of what sample environments and experimental techniques have been developed for soft matter research on neutron scattering beamlines with an emphasis, but not exclusively, on time-resolved small angle neutron scattering experiments, where one wants to follow the evolution of structure from an equilibrium state to one that is out of equilibrium, and vice versa.

2. Time Resolution

From a technical point of view, the raw neutron flux that a beamline can produce is an important parameter in defining the time-resolution that is achievable, but it is certainly not the only relevant instrumental parameter. Intrinsic background in the scattering pattern arising from the instrument, scattering contrast and sample chemical composition, sample size, detector efficiency, and associated electronics all play a role. Although the technical differences between experiments on a pulsed neutron source vs. a continuous source must be considered, in some ways they are less relevant than the type of sample environments used to carry out experiments, which can have a profound effect on the quality of the resultant data.

From a practical point of view, the time-scale over which the sample transforms from one physical state to another is the most important. However, equally important is how uniformly the sample can be perturbed. For instance, in a temperature jump experiment, temperature gradients are bound to develop. With the relatively large beam sizes produced by neutron scattering beamlines, one is detecting structural information over a thermal range. The faster the temperature jump, the larger the gradients, and the more complicated the data interpretation becomes.

The data quality per frame in a time-resolved experiment depends on the data collection time, which in turn depends on the progress in time of the process being studied. For different applications, different degrees of statistical accuracy are required. For example, if the goal is to perform a detailed structural analysis, the statistical quality of the data has to be very high. However, if one is satisfied with understanding how the radius of gyration of a particle in solution or the growth of a lamellar peak in a polymer crystallization experiment changes with time, the counting statistics requirements can be lowered and the experiments can be performed in a much more expeditious manner.

With the above points in mind, one can conclude that there is no definitive answer as to how fast a process one can resolve in a neutron scattering experiment. However, one should determine how fast a reasonable quality dataset can be acquired in order to address the problem and then assess whether or not the pertinent experiment is feasible to perform. New 'event mode' data acquisition will to some degree allow experimenters to have more leeway to carry out the experiment and afterwards decide on which time-resolution is feasible. 'Event mode' data acquisition involves retaining the position and time of each neutron detected, in contrast to traditional histogramming of detector data. Doing so makes it possible to parse a single data set collected into arbitrary time bins after data collection, which is extremely valuable for studies of time-dependent processes. It improves upon fast-frame data collection (i.e., a series of snapshots taken in quick succession) because it is not limited by the performance of the detector read-out, such as might be encountered with detectors used at synchrotrons.

For time-resolved experiments where a trigger is used to initiate a non-reversible process and the data are collected in individual time frames, this capability was available in the early 1990s with time-resolutions of between 2 and 3 min/frame to study late stage spinodal decomposition in polybutadiene-polyisoprene blends, where one of the blocks was deuterated [4]. Similar time resolutions have been reported for polystyrene-polyisoprene diblock-copolymers [5]. Nearly a decade later, a researcher studying micelle-to-vesicle transformations in D_2O was surprised about what was feasible and wrote: 'A measurement time of 60 s already results in astonishingly good statistics, although the total surfactant concentration is less than 1 mg/mL to avoid interparticle interaction effects' [6]. A decade later, in an experiment on the growth of mesoporous silica nanoparticles, using deuterium contrast variation, a time-resolution of 10 s/frame was reported. See Figure 1 [7].

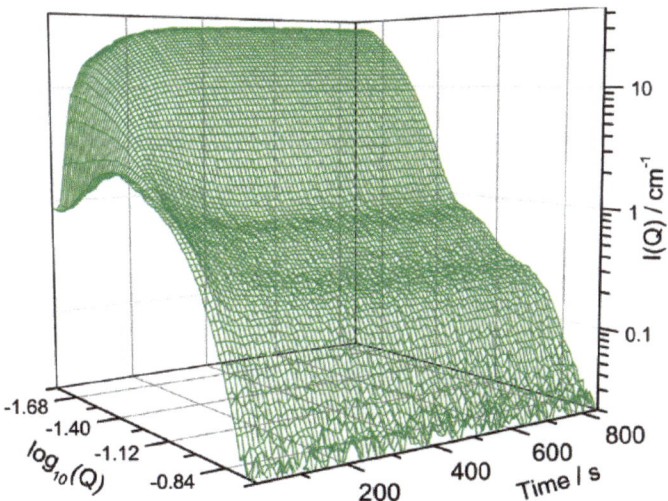

Figure 1. Time-resolved SANS data from the surfactant templated formation of mesoporous silica nanoparticles. This is one of the fastest neutron scattering experiments reported so far. A time-resolution of 10 s/frame rendered analyzable data (reproduced with permission from [7], American Chemical Society, 2012).

One may get the impression that every 10 years one gains a factor of 10 in time-resolution, but in reality, it demonstrates how strongly time resolution depends on the system being studied. Even though progress in instrumentation has been considerable, the quality of samples, to a great extent, still determines the achievable time-resolution. For example, in the case of a lysozyme crystallization experiment carried out in 1995, a time resolution of hours/frame was mentioned [8]. Similarly, 10 min/frame was reported for the growth kinetics of lipid-based nano-discs to unilamellar vesicles [9]. When not making full use of selective deuteration, one can still expect 20 min/frame of low statistical data quality in a demixing experiment of incompatible crude oils [10].

Improving the achievable time resolution by increasing the neutron flux is a very expensive process. The European Spallation Source (ESS) being constructed in Lund, Sweden and the proposed Second Target Station (STS) of the Spallation Neutron Source (SNS) at Oak Ridge National Laboratory in the USA will be the next generation of neutron sources offering increased neutron fluxes. However, both will require some years and large budgets until they are completed and ready for experiments. In the meantime, there are experimental methods currently available that significantly improve time-resolved experiments.

For experiments where cycling is an integral part of the study (i.e., oscillatory shear), or the periodic deformation used in simulations of the behavior of rubber tires, one can use a series of repeated short time frames and then merge the data series on-line. In principle, if the sample survives, one can collect for extended periods of time to obtain the required statistics and thus obtain a very high time-resolution. This method has been used for large oscillatory shear measurements (LAOS) on triblock copolymer micelles [11] and the effects of start-up and cessation of flow [12] using a commercial stress-controlled rheometer in the Couette geometry, where the rheometer generates an analog I/O signal that can be used to synchronize with the neutron scattering data acquisition system. In the case of the these strongly scattering systems, 300 cycles were sufficient to achieve 100 ms time-resolved data.

If an experiment that is not intrinsically cyclic can be accurately repeated, then one can follow the same procedure. Such a strategy can work if the sample can be recycled or if the synthesis/preparation is not too costly or time-consuming. Whether this is practically feasible depends on the length of the interval between experiments. In a temperature quench or pressure jump experiment, the dead time would be the required time to heat the sample up again before the next quench or to release/built up the pressure. A recent example of this is a series of pressure jumps that were used to study the phase transition kinetics of smart N-n-propylacrylamide microgels [13]. The jumps (200-40-200 bar) were performed with a frequency of approximately 23 Hz and repeated 5400 times. A time-resolution of 10 ms/frame was obtained (2 Hz frequency) over a period of 1.5 h of experimentation.

Another approach used to obtain good quality time resolution data of a continuous process is to translate the distance from the point where the perturbation takes place (e.g., mixing, extrusion, temperature change) to the point where the neutron beam interacts with the sample and translating this into time. By varying the time between the initiation point and the beam intercept, one can construct a timeline. An excellent example of this method is the (X-ray) study of polyolefin crystallization using an extruder [14]. Such experiments could easily be used on neutron beamlines as well. This translation method in combination with SANS was applied in studies of phase-separating systems, such as the cellulose nitrate/methyl acetate/isopropanol/deuterium oxide mixture, where the sample environment mimics the relevant conditions for industrial casting evaporative processes. See Figure 2 [15].

Better resolved time data are possible with the TISANE technique [16], and the first condensed matter results using this technique were reported in 2006 [17]. The technique can either be used on a pulsed neutron source or a steady-state source SANS beamline equipped with a timing chopper. Since this technique also requires periodic modulation of the sample, it therefore has a somewhat limited range of applications. Even though the TISANE technique offers great potential for the acquisition of very high-quality time-resolved data, it appears it is available to facility users, and it remains to be utilized to its full potential for time-resolved studies. The TISANE technique can, however, be improved through the use of time-of-flight techniques and event-mode data collection [18]. This can push the boundaries to 0.1 ms/frame but, again, only for experiments where cyclic data collection is feasible, and the samples scatter sufficiently strongly.

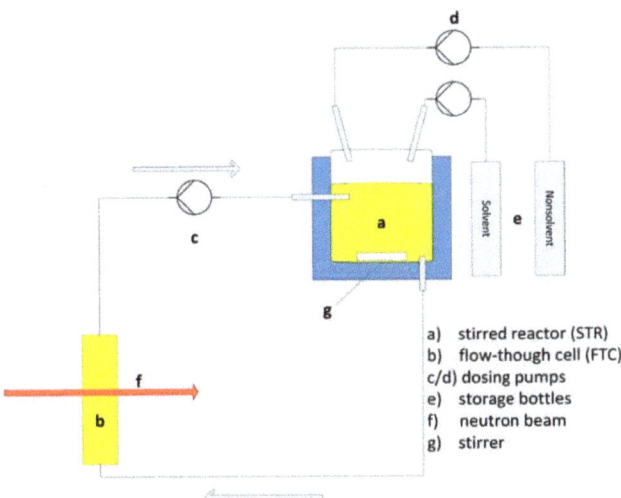

Figure 2. A flow-through cell used to study the liquid-liquid demixing process, which is the preliminary stage before casting the material to form a porous membrane. The degree of phase separation and the composition of the different phases determines the membrane structure to a large degree. The stirred reaction vessel is connected to the measuring cell, and the reacting mixture can be pumped through the neutron beam (reproduced with permission from [15], AIP Publishing, 2017).

We should point out that the above-mentioned examples are not intended to be a comprehensive list of what is feasible with time resolved neutron scattering experiments. Instead, the intention is to show that even though neutron beams are not as intense as X-ray beams, this is not a reason why one should not develop and carry out time-resolved studies. Although beyond the scope of this manuscript, one should also consider a change in mindset regarding time-resolved data. If the purpose of the experiment is to obtain insights into the evolution of the structure, instead of the accurate structural determination, a lower threshold for counting statistics data is acceptable as long as one properly analyzes the data and applies the appropriate error bars to the derived parameters.

3. Sample Environments

A sample environment can be a basic piece of equipment that is used to bring a sample to a steady state in order to measure the material properties and morphology or it can be used to bring the material out of equilibrium and in the experiment follow the structural evolution on the pathway to equilibrium. The latter can be of pure fundamental nature or a process mimicking an industrial process where several parameters like pressure, temperature, shear force, and tensile load are all varied simultaneously.

While many hard-condensed matter neutron scattering experiments use cryogenics, vacuum, and high magnetic fields, soft matter research has its own specialized sample environment requirements. Cryogenic conditions convert soft matter in general to hard matter, which is only in rare circumstances of interest in soft matter research.

Although neutron facilities historically were the first to develop sample environments for non-equilibrium studies, the advances in sample environments at synchrotron sources have in recent times taken place at a quicker pace than those at neutron sources. This does not imply that there is less demand for such developments, but it is more a reflection of the fact that it is more difficult to develop and implement novel sample environments at neutron sources, as a result of the size and intensities of the available neutron beams and the limited number of available instruments. While in the 1990s, X-ray beam sizes on the order of 300 µm were quite common, specialized beamlines currently can produce sub-micron size beams. Neutron beam sizes usually vary between 1 and 10 mm.

Traditionally, one key advantage that neutrons have over X-rays is their penetrating ability through different sample environment materials, making it possible for neutrons to study in a more facile manner materials under extreme conditions, such as elevated pressures, for example. However, the present generation of upgraded high energy storage rings generates high brilliance even in the photon energy range where the penetration is strongly enhanced and can be foreseen to encroach upon this neutron monopoly.

3.1. Temperature

Temperature is one of the basic parameters that can be varied during an experiment. This was already remarked upon by André Guinier in the proceedings of the first small angle scattering conference organized in 1965 [19]. For soft matter, research temperature control is important since the thermal history of a sample can dictate the material's ultimate properties. The main technical issue for neutron scattering is that neutron beam sizes are relatively large, making it more difficult to minimize sample environment temperature gradients. In the case of sample environments for static experiments, there are good commercial solutions to this problem. However, when there are special requirements, like the need to rotate the sample during a crystallization experiment, where the possibility of sedimentation is a real possibility, temperature control becomes somewhat more complicated, but in the range of between -10 and $80\,^{\circ}\mathrm{C}$, there are solutions [20,21]. In these cases, temperature control can be achieved using halogen light bulbs as the heating source and an infra-red sensor as thermal sensor, with an accuracy of approximately $\pm 2\,^{\circ}\mathrm{C}$. Also, accurate control can be achieved by using water baths.

For soft matter, however, the sample's thermal history can determine its final morphology and it is thus important to accurately control both heating and, especially, cooling rates. As such, a logical development is to combine neutron scattering with differential scanning calorimetry (DSC), a development that was implemented at X-ray facilities in 1995 [22,23]. The engineering difficulties due to the requirement to have access to the sample with larger neutron beams is evident since it took until 2014 before this was implemented on neutron beamlines [24] with an instrument that can operate in the temperature range $-150\,^{\circ}\mathrm{C}$ to $500\,^{\circ}\mathrm{C}$.

The larger beam size also implies that for time-resolved experiments involving temperature jumps, the situation is more complicated due to the low thermal conductivity of the samples. Fast quench rates, combined with the larger sample sizes required for neutron experiments, increases the problem of thermal gradients over the sample so that one does not obtain a realistic correlation between temperature and structure at a given temperature, but over a bandwidth of temperatures. However, compared to X-rays, there is the advantage that metallic window materials can be used which, when connected with the heating/cooling elements, can mitigate the gradient problem to some degree.

For jumps at higher temperatures, one could contemplate the use of microwave radiation [25], which can provide more uniform heating of the sample. However, for quenches to a lower temperature, material thermal conductivity remains the rate-limiting parameter. For liquids, one can perform temperature jump (T-jump) experiments by injecting the fluid into a pre-heated/cooled cell using, for instance, a syringe pump [26]. A temperature jump device specifically designed to operate on a SANS beamline, using two furnaces between which the sample is shuttled, can operate in the range 150–600 K with heating rates up to 19 K/s and cooling rates 11 K/s [27]. One can also adapt a stop-flow cell to operate as a T-jump device. This was done to understand the kinetics of the collapse, transition, and subsequent cluster formation in a micellar solution of P(S-b-NIPAM-b-S) [28]. See Figure 3. Approximately 100 s were required for a modest temperature jump of $6\,^{\circ}\mathrm{C}$.

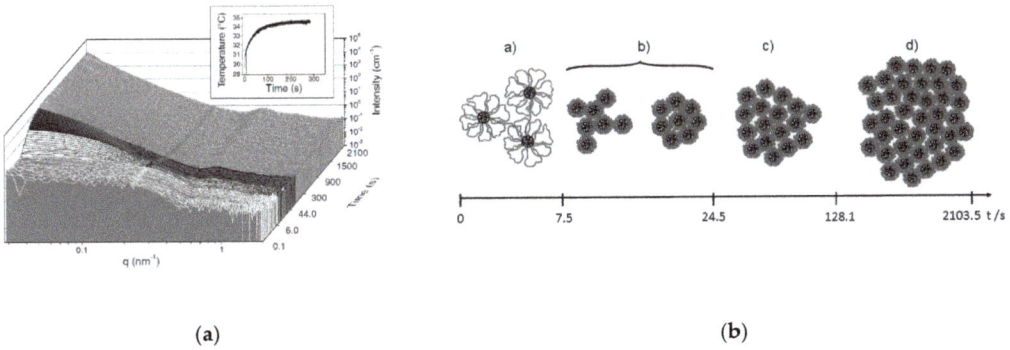

Figure 3. (a) Left hand panel, SANS curves obtained during a temperature jump on a micellar solution of P(S-b-NIPAM-b-S), from 29 to 35 °C, where the first structural changes were observed 7 s after initiation of the experiment. (b) The right-hand panel shows the structural models determined from the data, where micelles collapse and subsequently aggregate into clusters (reproduced with permission from [28], Wiley, 2012).

The latter example shows that it is not always necessary to use world record T-jump speeds to obtain the desired information with respect to the scientific question to be answered.

3.2. Pressure

Pressure is the thermodynamic parameter that is relatively easy to change homogeneously. For the larger sample volumes routinely used in neutron scattering, this allows the reduction/elimination of large pressure gradients over the sample, allowing for a better controlled experiment producing more robust results. Most high-pressure experiments use hydrostatic pressure, which is not directional. An example of uni-axial pressure exerted on a sample in combination with a SANS experiment can be found in soil- or geo-mechanics [29]. The pore size distribution in bentonite clay was investigated as a function of pressure whilst simultaneously loading the cell with dry CO_2 and water in order to elucidate the role that intercalation of these two components play in the complicated pore-pressure diagram. Pressures up to 10 MPa at ambient temperature were applied.

An example where hydrostatic pressure was applied was in research to determine if CO_2 could be stored in porous shales. Here, CO_2 was not only the object of study but also the neutron contrast variation medium [30]. Pressures of 25–40 bar were achieved over a temperature range of 20–60 °C. This type of study can be relevant to soft matter where, for example, the porosity of a material must be characterized to understand the impact of processing conditions.

Not only does the static characterization of pores attract attention, but also the combination of hydrostatic and uniaxial pressure is relevant for understanding flow in porous materials [31]. In the fracking of oil-containing shales, water is used as the pressurizing agent. In the case of neutron scattering experiments, this allows one to exploit the difference in scattering length between light and heavy water, in addition to the natural air-matrix scattering length difference. Uniaxial stresses, σ_{ax}, in combination with the hydrostatic pressure inside the pores, p, allows to generate an effective stress $\sigma_e = \sigma_{ax}-p$ of 600 Bar.

An example of the effects of hydrostatic pressure is surfactant solutions of SDS in D_2O-forming microemulsions, as shown in Figure 4. Here, a static pressure was applied and the effect of pressure is the continuous transformation of spherical particles to elongated micelles [32]. The pressure cell used was capable of a maximum pressure of 10 kbar and a maximum temperature of 80 °C.

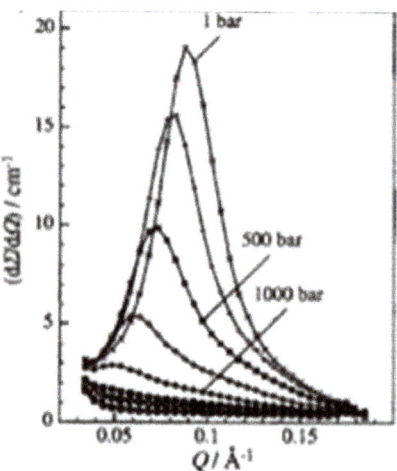

Figure 4. The effects of hydrostatic pressure on a micellar SDS solution as observed in a SANS experiment. At increased pressures, there is a continuous transformation from spherical to elongated micelles as evidenced by the gradual peak shift to lower angle. Figure provided by author [32].

A detailed design for a temperature-controlled high-pressure cell developed for biological applications, specifically protein denaturation studies, was described by Teixeira et al. [33]. The sample cell was capable of a low temperature of −18 °C with a pressure of 2 kbar. Because of the cell's large thermal mass, which is required for mechanical strength at these elevated pressures, it required 1 hour for sample equilibration. When working at sub-zero temperatures, one stands the risk of ice crystal formation. Down to about −20 °C, ice formation can be inhibited by anti-freeze agents, but this will alter the biologically relevant environment of the proteins. By using pressure, the freezing point can be lowered, and the use of anti-freeze agents is avoided. There are several other sample cell designs available [34].

Pressure jumps can perturb a sample over the whole volume exposed to the neutron beam without any pressure gradients and can be relatively simply and accurately repeated. This makes it possible to perform such experiments with a reasonably high time resolution using the 'cyclic' procedure mentioned in the section on temperature jumps.

An example is the phase transition from swollen chains to polymer mesoglobules of an aqueous solution of poly(N-isopropylacrylamide) with pressure jumps across the coexistence line [35]. This was investigated with kinetic small-angle neutron scattering with a 50 ms time resolution. The pressure jumps were on the order of $\Delta p \approx 200$ bar. The interesting variation here is that the data acquisition time per frame was not constant but starting with a time frame length of 0.05 s that was elongated by a factor of 1.1× for each frame, thus anticipating that events evolved fastest immediately after the pressure jump. Obviously, this has some consequences regarding the statistical quality of the data, but the experiment only had to be repeated five times. This is another area where 'event mode' data collection will allow the experimenter more freedom to postmortem decide how to time-slice the data set.

A similar development was used to study the volume phase transition kinetics of N-n-propylacrylamide microgels [13]. A jump sequence of 200 → 40 → 200 bar was performed using a time frame length of 5 ms over a total time period of 350 ms. However, in this case, the experiment had to be repeated 5400 times before satisfactory data sets could be obtained.

It should be noted that pressure jump experiments, as those described above, are not only interesting for fundamental research purposes, but also understanding industrial processes. However, to do so, the apparatus should be capable of jumping both pressure and temperature simultaneously, preferably with the application of a shear force [36,37]. This type of experiment has been implemented for X-ray scattering, but it is probably still beyond the limits of what is feasible for neutron scattering due to beam size limitation and neutron flux.

3.3. Shear and Rheology

The flow of materials is a research area of interest to both the fundamental as well as the applied/industrial research communities, and one that lends itself to neutron scattering experiments. The cell walls of conventional Couette cells commonly used in neutron scattering experiments can be easily penetrated by neutrons and the amount of material required is small, which keeps deuteration costs to a minimum. Couette cells have a long history of commercial development and as such, are a mature technology able to keep a stable flow over the entire time required for the acquisition of data with good signal to noise ratio; one of the first mentions of the use of a Couette cell in a scattering experiment was in 1984 [38].

Rheology, however, is a complicated subject in which the observed components of a flow profile depend on the configuration of the cell or in which direction one probes the flow profile within the cell. In Figure 5, the different flow planes are defined and further discussions about the terminology can be found in an extensive review [39]. For the directions where one can probe the desired flow field whilst keeping the Couette cell axis vertical, i.e., the 1–3 and 2–3, radial and tangential directions, respectively, it is feasible to perform quantitative rheological experiments with neutron scattering [40,41]. However, a cell for the 1–2 direction, where one has to align the neutron beam in the gap parallel between the two Couette cylinders, has proven more difficult to design [42]. With the Couette cylinder axis horizontal, the material must be confined, otherwise gravity will force low viscosity materials out of the cell.

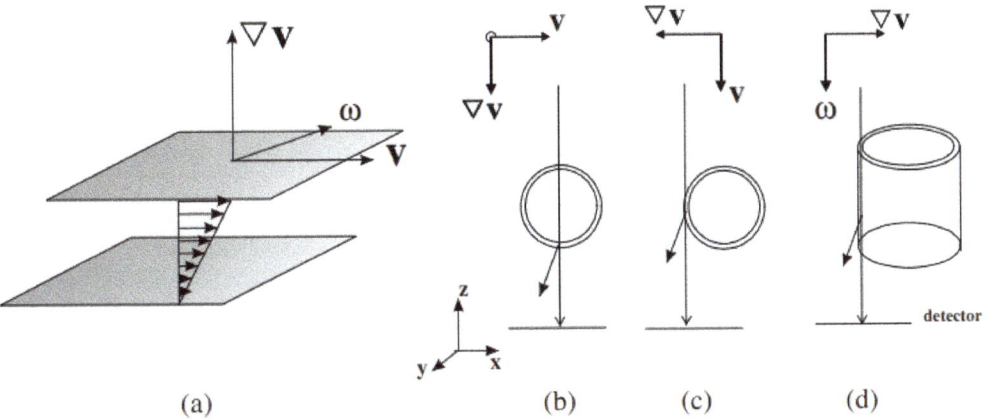

Figure 5. The different configurations in which a Couette type shear cell can be used. The different flow fields are indicated with v indicating the flow direction, ∇v the flow gradients, and ω the vorticity direction (**a**). In neutron scattering experiments, the 1–3 (**b**) and 2–3 (**c**) directions can be probed with a cell having a vertical rotation axis. On the other hand, the 1–2 (**d**) direction requires a cell with a horizontal rotation axis. Figure provided by J. Vermant [43].

For the conventional Couette geometry where the cylinder axis is vertical, one can rely on commercially available rheometers. Since the topic of shear and rheology is rather popular, there have been numerous in-depth reviews on the subject [39]. These reviews include oversights of other available shear geometries that have been used on X-ray and neutron beamlines. Measuring the axis torque allows quantitative measurements of shear stress simultaneously with macroscopic strain gradients imposed by the frequency of rotation and microscopic deformation observed in the SANS data [42]. As mentioned, for most neutron scattering experiments there are commercially available instruments, where the replacement of the commercial Couette cylinders by cylinders made out of neutron transparent materials like quartz and titanium can be straightforward.

The literature contains a wide range of subjects that have been studied using Rheo-SANS methods. They range from investigations into the shear banding mechanism in a fluid comprised of cetyltrimethylammonium bromide wormlike micelles [44], to detailed studies of polymer chain conformations composed of polyethylene-polypropylene blends with Si particles nanocomposites [45], to the shear-induced formation of rod-like entities in metallo-supramolecular gels based on a multitopic cyclam bis-terpyridine [46]. The Couette geometry can also be used to probe material relaxation properties by using an oscillatory flow protocol, which is mentioned in the section on time-resolution.

The 1–3 and 2–3 Couette geometries are now fairly standard for neutron scattering experiments. Progress in instrumentation in the coming years will most likely be incremental and most developments will focus on the analysis of neutron-rheology data sets [47]. Some developments were recently reported on the implementation of the technically more challenging case, where one wants to probe the 1–2 plane [42,48] by directing the neutron beam through the gap between the two cylinders that make up the Couette cell. With a small enough beam, it is possible to probe the different flow profiles as a function of the position within the gap [48]. Such spatial mapping is not commonly associated with neutron scattering since the neutron beams tend to be rather large and can only be reduced in size by sacrificing intensity. For stable flow rheological applications, this is a problem only limited by the available beam time and neutron flux, and not by any technical aspects. From an experimental point of view, it should be remarked that it is often necessary to obtain scattering data along all three shear planes [49]. Unfortunately, this requires the use of two different Couette cells capable of covering the same temperature and shear range. Designs for a cell that can accommodate experiments in all three shear directions are available, but rare [42], and still require substantial reconfiguration when changing between geometries.

In addition to the above-mentioned shear flow geometries, there is the less popular but useful plate-plate geometry [50] used in a range of SANS experimentation, such as studies on the self-assembly of nanofibers composed of fibronectin mimetic peptide-amphiphiles [51].

The design for a cone-plate geometry instrument for use in reflectometry is discussed in the framework of test experiments involving polystyrene/deuterated polystyrene, with the flat bottom plate modified to be transparent to neutrons [52]. A modified commercial instrument, for example, was used for the study of asphaltenes in the reflectometry configuration [53]. However, placing the neutron beam in the gap between the cone and plate, like what has been successfully done with X-rays, requires that very tight neutron beam collimation be used, which impacts data rate and quality. An interesting development is the use of Grazing Incidence SANS or GISANS. The feasibility of this technique in combination with a cone-plate rheometer was demonstrated in experiments on bola-amphiphilic arginine-coated peptide nanotubes [54]. However, data collection took about 5 h per sample, indicating that this method still requires further development.

The more difficult problem of how one goes about studying elongational flow appears to still be out of reach of real time neutron scattering experiments. The methodology of flash freezing a material that has been subjected to elongational/extensional flow off-line and then using the frozen samples to carry out a static measurement was reported in 1990 [55], and the method was still regarded as recently as 2016 as being the best method to study dendritic polymer blends with linear chains [56]. Here, the main issue remains that the time-resolution of the physical process does not match the time-resolution that can be reasonably achieved on SANS beamlines.

Whilst the examples being discussed above fall under the 'drag flow' category, there are also configurations that allow 'pressure flows' to be studied [57]. Capillary- or Poiseuille-flow and slit-flow are examples of pressure flow. This type of flow is important to understand the fundamental behavior of chain molecules in conditions relevant for industrial processing. For instance, in injection molding, the predominant physical processes are Poiseuille flow and thermal quenches. Not only is flow in an unrestricted path relevant, but studies of the behavior of the material as it flows around obstructions and restrictions are also required. Such information provides firstly, insights into material properties, and secondly, these results can be used to validate computational models as to how entangled polymers behave under elongational flow or when encountering a change in flow geometry. In other words, such physical results provide predictions on how to process materials [58,59]. When run in an 'open' configuration, these set-ups require considerable amounts of material, which with deuterated samples tends to be prohibitively expensive. Hence, efforts have gone into recirculating flow cells, which require relatively small amounts (~200 g) of material [60–63]. This cell is equipped with non-birefringent sapphire windows that are strong enough to resist pressures of up to 10 MPa, and also allow for birefringence measurements to be carried out, thus creating a set-up which can deliver structural and orientational information over a wide range of length scales. See Figure 6.

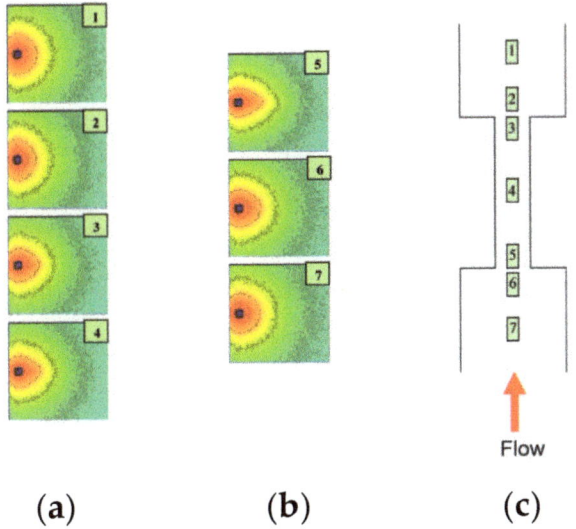

Figure 6. 2D scattering patterns from a material encountering an obstruction using a pressure-driven slot flow cell (**a**,**b**). The neutron beam is directed at right angles with respect to the flow direction (**c**). The numbered boxes correspond to the different SANS patterns. The anisotropy is due to the deformation of the backbone of the comb-shaped polymer seed in this experiment (reproduced with permission from [63], ROYAL SOCIETY OF CHEMISTRY, 2009).

A recent design based on developments in microfluidics [64], which itself was based upon a larger scale design by G.I. Taylor from 1934, allows for a variety of different flows to be applied within a single sample environment. The Fluidic Four Roll Mill, so named because it uses four rolling cylinders [65], is a device consisting of 4×2 orthogonal channels in which flow patterns can be generated by opening/closing a series of valves [66]. See Figure 7.

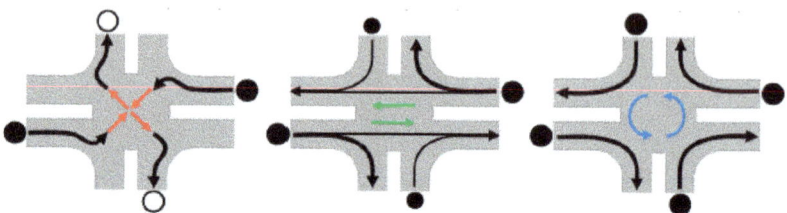

Figure 7. A schematic of the fluidic Four Roll Mill device in which different flows can be generated. The flows can be created by applying a pressure flow in the channels marked with closed dots and opening the valves in the channels marked with the open dots. The neutron beam is directed along the white path between the flow channels with a diameter of 1 mm. The flow channels have dimensions of 2×3 mm^2 (reproduced with permission from [66], Springer, Nature, 2018).

3.4. Mechanical Deformation

Uniaxial mechanical deformation studies on soft materials are relatively easy to implement and are extensively used by academics and in industrial materials testing laboratories. Several small-scale deformation stages are commercially available, but these are in general, only suited for qualitative tensile data involving local molecular conformations. To obtain quantitative tensile results simultaneously with the neutron scattering data, one has to make sure that the sample is uniform over its entire volume and not only at the point of intercept with the neutron beam. With the exception of ambient or near ambient temperatures, one encounters the issue of how to uniformly heat the sample. This obviously can be achieved by placing the entire deformation load frame in a temperature-controlled chamber, but in doing so, any temperature changes will be slow to equilibrate, thus limiting the technique's application with neutron scattering instrumentation.

Since polymeric materials exhibit a wide variety of materials properties, the type of required load frame depends on the maximum tensile force and elongation required. Elastic rubber samples require little stress but a high strain, i.e., elongation [67], whilst the reverse is true for High Density Polyethylene (HDPE) [68]. For such samples, the required mechanical performance (i.e., stresses between 10 and 200 N) required of the load frames is easily covered by commercially available equipment from a variety of suppliers.

Early on, mechanical deformation experiments were performed in a quasi-static fashion, i.e., deform-hold-collect data. This deformation protocol works well for some materials where the molecular relaxation times are long. Otherwise one has to revert to continuous stretch protocols where the data collection time can be problematic. This is often the situation when the materials are unclamped and allowed to molecularly relax [69]. Relaxation of isotopic blends of linear polyethylene and ethylene copolymers with butyl and hexyl was studied, but due to the relatively small amount of material in the neutron beam (intrinsic to this type of deformation experiments), data could only be obtained at 30 min time intervals. An extensive discussion on how the anisotropy manifests itself in scattering patterns, due to the molecular and domain orientation induced by applying the deformation, can be found in experiments of elastomeric polypropylene [70].

A method to increase the information content of on-line deformation experiments is the use of Digital Image Correlation (DIC) [71]. This optical method is easy to install on a neutron scattering instrument and allows one to probe the homogeneity of the deformation process over the entire sample and also provide macroscopic material parameters. See Figure 8. When combined with neutron scattering experiments, one obtains a more complete picture of the material's deformation behavior. Simultaneous DIC furthermore supplies additional validation of the reliability of the experimental results since mechanical problems, such as slippage at clamps, will be noticed earlier.

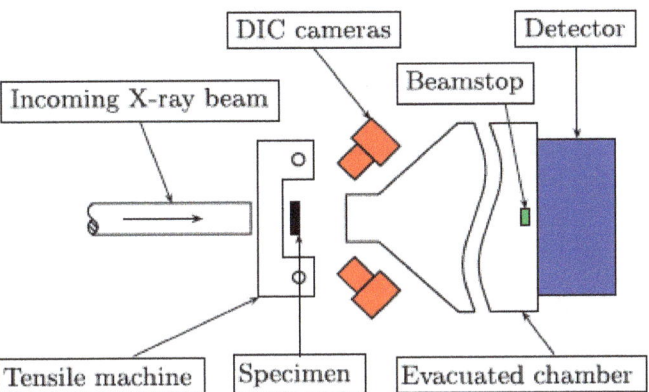

Figure 8. An experimental X-ray/neutron scattering set-up incorporating optical digital image correlation. This combination of experiments can increase considerably the information content of experiments and is relatively simple to implement. Figure made available by authors [71].

Although homogeneous temperature control on samples that require large elongation is somewhat more complicated to achieve, other parameters like humidity can be relatively easy controlled. This requires the construction of a solid chamber around the moving parts. An example of this is the stress-induced long range ordering in spider silk [72]. Here a combination of finite element modelling and SANS data were used to elucidate the role that crystalline domains play in the ordering of silk when extended. The crystalline domains are less susceptible to water uptake compared to the amorphous domains, and this physical characteristic allows for contrast variation to be used. By placing the tensile load frame in a 100% D_2O relative humidity environment, it was possible to selectively deuterate the amorphous parts of the silk.

One of the issues that one might encounter due to beam intensity and the samples becoming thinner when elongated is that the statistical quality of the scattering patterns is rather poor. One might be tempted to use a stop-start technique where the material is not continuously stretched but instead, after a short stretch, a pause is taken to allow data to be collected. Of course, this has the problem that the sample is allowed to relax and the results will not necessarily be the same as with a continuous stretch. The differences between the results obtained with these two approaches when deforming polypropylene are discussed in the literature [73].

3.5. Stop Flow/Chemistry On-Line

To be able to follow structure formation due to chemical reactions requires reaction cells that are resistant to the chemistry being carried out and also capable of attaining the desired temperature and pressure range. In the case of low viscosity solutions, the reaction products can sediment out and therefore change the composition of the sample as 'seen' by the probe beam [74]. This can be overcome by the use of a tumbling cell, which rotates the sample solution around the beam axis, preventing sedimentation. These cells are routinely used in all neutron scattering facilities. As an alternative, a cell of sufficient thickness with a mechanical stirrer can also be used. As with any equipment that contains moving parts, this leads to a more complicated temperature control system. One design that contains an array of four tumbling cells, makes use of air cooling/heating, is capable of a limited temperature range of 10–50 °C, and is suitable for slow reaction kinetics without thermal variations [20]. By placing the equipment on a translation stage and triggering the chemical reactions at different times, it is possible to alternate between the samples and map out the kinetics over extended time scales, also making effective use of the allocated beamtime.

For faster experiments, sedimentation is less of a factor and here one can consider the use of conventional stop-flow cells. Due to the required beam size, the sample volume will be rather large, and thus there are constraints on the achievable mixing- and dead-times, but workable solutions do exist. A review on the use of stop flow cells in SAXS and SANS experiments was written by Isabelle Grillo [75], who pioneered such experiments on SANS beamlines. A practical example of such experiments is the formation of microgels due to the precipitation and subsequent polymerization of N-Isopropylacrylamide [76]. Here the early stage of gel formation was the point of interest, which is often the most difficult part since structural changes tend to develop faster than in later stages. A commercially available stop-flow cell was used at a fixed temperature. Data were collected over a period of 20 min with a time framing rate of 5 s/frame. However, in order to obtain sufficient quality data, the experiments were repeated three to four times and the results were averaged. The data yielded the total particle volume and the number density of particles. A similar home-built device was used to report on cationic liposomes complexes with DNA. However, instead of temperature, a pH jump was induced by the stopped flow to trigger the reaction [77].

By using continuous flow methods, a larger variety of experiments is feasible even though much larger quantities of materials are required. In some cases, it is even feasible to use a recirculating flow and gradually change the conditions without incurring the penalty of requiring large volumes of sample. This is especially relevant when dealing with deuterated materials. A purposely designed sample environment meeting these requirements was reported. See Figure 9 [78].

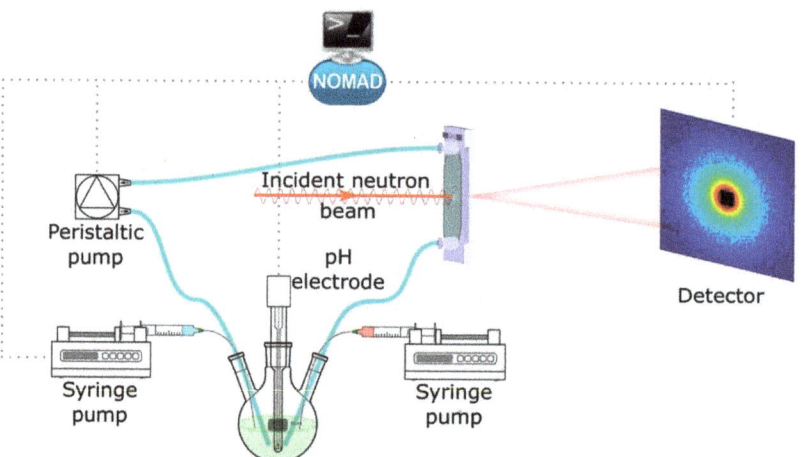

Figure 9. Schematic of an experimental set-up that allows structure formation due to chemical reactions to be followed in real time by SANS. The system consists of a reaction vessel where the reactions can be initiated using remotely controlled syringe pumps. A peristaltic pump is used to bring the mixture to the measuring cell. The system has the option to be used in open or closed loop configuration. In a closed loop system, one can, for instance, observe the effects of a gradual change in pH. Here, 'Nomad' is the name of the central control computer (reproduced with permission from [78], Springer Nature, 2018).

For test experiments, aqueous solutions of polyoxyethylene alkylether carboxylic acids, a class of surfactants with strong pH-responsive properties, were used. During the experiments, the pH was changed by titration. As the authors remarked, these time-resolved experiments, down to 1 s/frame, were made feasible in recent years by more efficient detectors and increased neutron flux thanks to improvements in neutron guide technology.

Pressurized CO_2 gas can be used as a neutron scattering contrast agent in porous materials like lignite and shale [79], and SANS/USANS experiments are part of the toolset that can be used to determine pore distribution size and morphology for pores that are accessible to the CO_2. However, when introduced in its supercritical state ($ScCO_2$), it is a good solvent, primarily for nonpolar low-molecular-weight compounds. For polar higher molecular weight compounds, it is a rather poor solvent. Thanks to the fairly benign conditions at which criticality occurs (7.39 MPa, 31.1 °C), $ScCO_2$ is a fairly attractive option for performing chemical synthesis of surfactants, polymers, and biomaterials [80]. On-line neutron scattering investigations of poly(dimethylsiloxane) polymer-$ScCO_2$ solvent phase diagrams were performed and fundamental applications demonstrated [81]. Time-resolved scattering experiments where the $ScCO_2$ to CO_2 transition was used as a foaming agent to create micro and nano foams allowed the kinetics of foam formation to be studied as a function of pressure and pressure modulation [82]. The use of fluorinated surfactants in CO_2 microemulsions is widespread but not desired for environmental reasons. In a systematic pressure and contrast variation study, it was shown that partial substitution of $ScCO_2$ by cyclohexane reduced the required amount of fluorinated surfactants considerably [83]. Contrast variation through the use of D_2O exchange was crucial in this case and the only way this kind of structural information could be obtained.

Performance of on-line chemical synthesis experiments has so far not been widely explored, although it appears that neutron scattering could play an important role, as the above example has shown. When used for structure forming on-line chemical processing, temperature control and homogenization of the reaction mixture is of the utmost importance. The designs for such cells used in on-line experiments should take this into consideration, as well in allowing for the possibility of siphoning off small amounts of the reaction mixture (for off-line chemical analysis) during the course of the experiments without perturbing the pressure/temperature conditions [84].

The use of on-line size-exclusion chromatography to deliver well-defined and non-aged samples for scattering experiments has become feasible, although here one has to keep in mind the restrictions on the amount of available material. This method requires beamlines that can deliver high intensity and small beam sizes. However, when these conditions are met, the experimental accuracy can be considerably improved. For example, proteins that have a tendency to aggerate in solution can still be studied in their non-aggregated state [85].

3.6. Electromagnetic Fields

Sample environments generating electric fields have not been extensively developed for use at neutron beamlines. AC fields that can be used to (partially) align relatively short rigid polymeric molecules in solution in order to perform fiber diffraction experiments have been used for biological molecules. The main problem with this approach is that the frequencies used should be such that the sample does not heat up and that the fields are high enough that they are not shielded by the salts in the buffer solution [86].

The main problem with using static electric fields is that the voltage required as function of sample thickness is considerable. When using too high a field, one risks electric breakdown damaging the sample. To induce domain alignment in thin films of symmetric diblock-copolymer of polystyrene and poly(methylmethacrylate), (PS-b-PMMA) fields of 40 V/μm were used in a sample cell consisting of an aluminumized Kapton film used as an electrode and a similar electrode that is isolated from the sample by a thin layer of poly(dimethylsiloxane) [87]. The alignment process could be followed by time-resolved SANS experiments. As to be expected, the interplay between the sample thickness and the interfacial interactions between the electrode and the sample plays an important role. For thin samples, where the interfacial interactions are most important, the structure can evolve over periods of hours. On the other hand, for thicker films, complete alignment takes only minutes. Static electric-field-induced deformation of bulk poly(styrene-block-isoprene) lamellar phases, swollen in toluene or tetrahydrofuran, has been studied with SAXS and with SANS. Capacitor cells with 3–5 mm electrode gaps and typically 5 mm beam path length were used to apply up to 60 kV, reaching electric fields up to 12 kV/mm [88]. See

Figure 10. Electric breakdown is a common limitation under these conditions, sufficient stability for the duration of experiments can however be reached with careful control of experimental conditions, paying very careful attention to drying polymer and solvent to remove any traces of water.

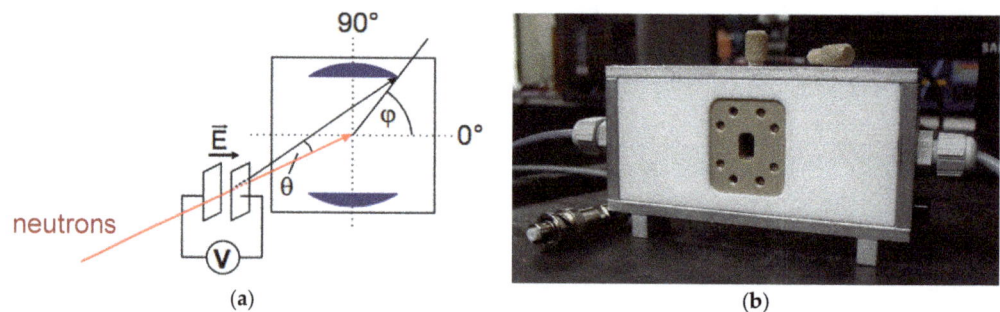

Figure 10. Electric field setup used for on-line SANS experiments [89]. Schematic drawing of experimental setup (**a**). Capacitor dedicated for small-angle neutron scattering experiments of polymer solutions (**b**). The capacitor consists of gold electrodes and its interior parts are mantled in Teflon.

Dielectrophoresis uses the intrinsic polarizability of objects/molecules to be manipulated in a gradient or AC electric field. When this method is applied in combination with SANS, it is possible to follow the development of molecular ordering due to different field strengths and frequencies. A model system, polystyrene particles in H_2O (diameter 195 nm and 530 nm), and an applied AC field formed colloidal crystals consistent with the space groups P6mm and C2mm [90]. A schematic overview of the cell is shown in Figure 11. The AC field was applied in transverse direction to the neutron beam and depending on the experiment, a range of $0 < \omega < 100$ kHz and $0.62 < E < 1.4$ kV was used with a cell thickness in the neutron beam direction of 130 microns. The cell thickness was limited in this design due to the requirement of exposing the sample to a reasonably homogeneous AC field; a thicker cell would contain field gradients.

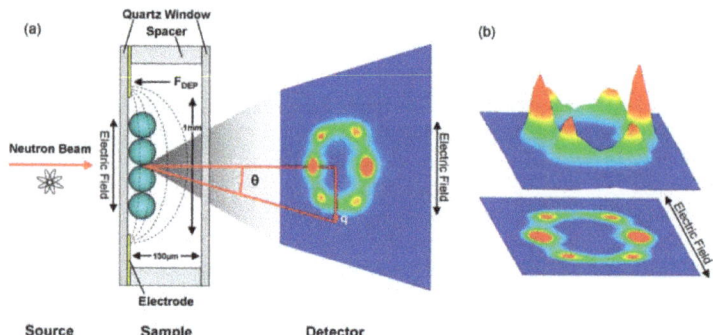

Figure 11. Top view of an electrophoresis cell used for on-line SANS experiments (panel (**a**)). The AC field is applied perpendicular to the beam direction and induces the formation of a colloidal crystal. Panel (**b**) shows a representative diffraction result. Figures not to scale (reproduced with permission from [90], The Royal Society of Chemistry, 2010).

Where the formation of colloidal crystals by the application of an AC electric field is maybe somewhat counterintuitive, the formation of oriented bundles of rather stiff fibrous molecules of the conjugated polymer poly(3-hexylthiophene) (P3HT), with a persistence length of several nm, is less surprising [91]. The authors are also describing parallel optical experiments where the data were Fourier transformed to provide an estimate of the degree of orientation.

One of the surprising findings is that on-line strong magnetic fields have so far rarely been used as a sample environment in soft matter-related neutron research. In most neutron scattering laboratories, there is an abundance of resistive and superconducting magnets available for research on magnetic materials. One would have expected that several potential user groups would have taken advantage of this opportunity and used these magnets for soft matter experiments as well. For soft matter, one clearly requires higher sample temperatures than the cryogenic temperatures most commonly used in magnet research and most often a configuration where neutron beam and magnetic field are at 90 degrees, thus requiring a split coil magnet over a simpler solenoid.

In liquid crystal and liquid crystalline polymer research, the study of material response to magnetic fields is common, but often only moderate fields (<2 T) are required. An example of this is the interplay between ordering and microphase separation of liquid-crystal polystyrene-poly(methyl methacrylate) block copolymers bearing a chiral biphenyl ester side group linked to the backbone by a dodecyloxy spacer [92]. The applied magnetic field causes the polymer backbone to partially orient. Experiments on another liquid crystalline block copolymer to map out the temperature-magnetic field phase diagram have also been reported [93]. When doped with lanthanides, lipid membranes can also be oriented [94].

Since the interaction of soft matter often is via diamagnetism, the required fields to be able to observe a relevant alignment interaction are >5 T, i.e., greater than the fields produced by permanent and electromagnets. The requirement to use superconducting magnets is a severe complication. Some static on-line X-ray experiments are published [95,96] and even some dynamic studies, where the sample was rotated inside the magnetic field and observed whilst rotating back [97,98]. For neutron experiments, occasionally an off-line magnet was used to produce aligned samples from fibrous biological macromolecules [99,100] and more recently the use of an on-line 8 T magnet to measure alignment by SANS of bicelles complexed with lanthanides was reported [101].

Transportable higher field on-line superconducting solenoids (17 T) have been used, but so far, no results on soft matter samples have been published [102]. For higher continuous fields, one has to resort to more permanent installations like Bitter magnets, very large superconductors, or hybrids of these two. Pulsed magnetic fields using capacitor banks are often transportable and split coils can reach up to 30 T [103–105] and solenoids up to 40 T [106], but the short pulses (msec) and the slow repetition rates (several shots/hour) make these magnets unrealistic options for soft matter research.

3.7. Light

The use of visible or UV light to modify samples on-line has found some applications in chemistry and in biological systems. An example is the use of UV light in a solution with a UV photo-destructible anionic sodium 4-hexylphenylazosulfonate (C6PAS) surfactant in combination with an inert surfactant hexaethylene glycol monododecyl ether (C12E6). Upon exposure to the UV light from a high-pressure Hg lamp, the C6PAS fell apart and released the content of its micelles to react with the inert surfactant. Hg lamps produce a high light intensity so care should be taken in controlling sample temperature [107]. Similar experiments on reversible phase separation systems using narrow bandwidth light sources have also been reported. Here two different UV wave lengths (350 and 450 nm respectively) were used to initiate phase separation and to reverse it [108]. Photorheological studies, where the rheological material properties are influenced by the exposure to UV light, have also been reported, but it should be noted that the UV irradiation was not performed on-line with the SANS experiments [109].

Visible light produced by a 'white' halogen lamp was used in SANS experiments to gain insight into the gelation process of the conjugated optoelectronic polymer poly(3-

hexylthiophene-2,5-diyl) (P3HT). Here the effect of the exposure to light was found to retard the growth of microstructures [110,111]. Although the authors mention that they monitored the light intensity, no information about differences in growth of microstructures as function of dose rates was provided.

The conformation of biological molecules involved in the photosynthesis process as a function of exposure to light is a research subject that is well-suited for study by SANS. Not only because of the advantages that contrast variation via deuteration can bring, but also due to the fact that one does not have to worry about the interactions of photoelectrons, which invariably are created in X-ray-based experiments. Here, the issue is not so much recognizable radiation damage, but instead, the less recognizable effects of the direct interaction of X-ray photoelectrons possibly causing conformational changes [112]. Using live cells that were, both on- and off-line, exposed to 'cool white LED light' with an intensity of ≈ 20 μmol of photons m^{-2} s^{-1}, the structural organization of membrane systems in cyanobacteria were investigated [113]. An illustrative example of the results is shown in Figure 12. It should also be mentioned that quasi-elastic neutron scattering experiments were performed under similar illumination conditions.

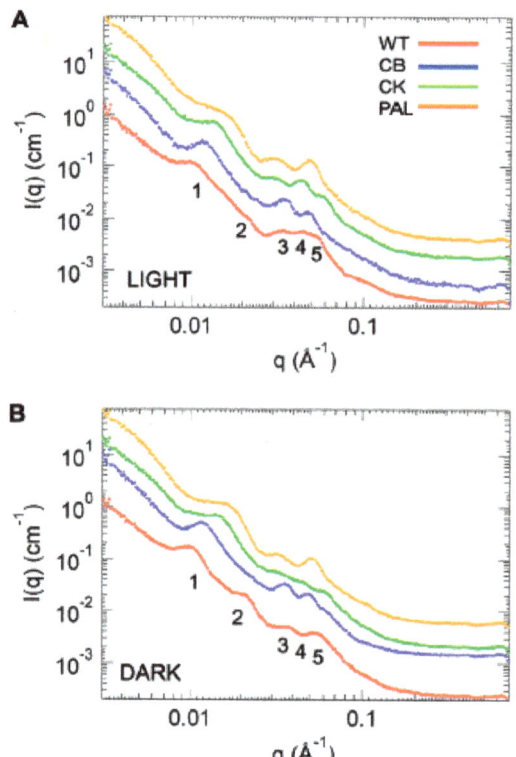

Figure 12. SANS data from the membranes of live cyanobacteria cells in both light and dark conditions. WT, CB, CK, and PAL indicate data from different cyanobacteria mutants. The panels marked (**A**,**B**) show data obtained under light and dark conditions, respectively. There are subtle differences apparent between the two illumination conditions. The numbers indicate the different peaks visible in the spectrum. For a full explanation, one is referred to the original manuscript. Figure adapted and provided by the authors [113].

Similar experiments, using SANS and Quasi Elastic Neutron Scattering (QENS), were used to determine the volumetric changes of the light sensitive biological pigment rhodopsin, although the experimental details with respect to the exact illumination conditions are somewhat vague in the manuscript [114].

Microwave radiation can be used for sample heating, but an alternative use is to influence chemical reactions, although the exact way that microwaves influence chemical reactions is not entirely clear. It can change the rate constants, but the reaction pathway may also change. Preliminary neutron scattering experiments with an on-line microwave generator were performed, but these experiments were inconclusive [115] and have not received much follow-up. This may be a missed opportunity since in a later work using X-ray scattering, it was shown that by adding microwave-interactive chemical species it was possible to perform targeted annealing of specific molecular regions [116]. Such a use of microwaves, in combination with the advantages that selective deuteration can impart, may allow one to gain insights in chemical pathways.

3.8. Container-Less Measurements

In some cases, contactless or container-less experiments are carried out to minimize the interaction of the sample with the sample cell or to reduce the background scattering due to window materials. This is particularly important for very high temperature experiments, where the sample–wall interaction can change the chemical composition of the sample [117]. So far, aerodynamic levitation, where a vertical gas stream keeps a droplet of material in the beam, is most commonly used to study materials at high temperature. Powerful lasers are used as a heating source [118]. For soft matter research, aerodynamic levitation appears to be less popular and acoustic levitation is used, instead [119]. A test experiment on the drying of lysozyme solution droplet has been reported [120]. Although this experiment was intentionally aimed at drying and thus increasing the protein concentration, it also highlights one of the method's shortcomings, namely solvent evaporation. The authors also noted that the strong 22 kHz soundwave, which helps to maintain internal mixing, also has the unwanted side effect of denaturing the protein.

Independent of which levitation method one uses, a drawback is that one can only levitate droplets that are smaller than the average neutron beam, thus reducing the scattering intensity as well as possibly introducing parasitic scatter from the sample-air interface. Even if using a jet of liquid shooting through the beam, one will still be restricted to small sample sizes since it is difficult to create large, stable laminar flows. Another ingenious approach to contact-free measurements that mitigates the problem of size mismatch between sample and neutron beam is to create a free liquid film by flowing a solution between two wires [121]. See Figure 13.

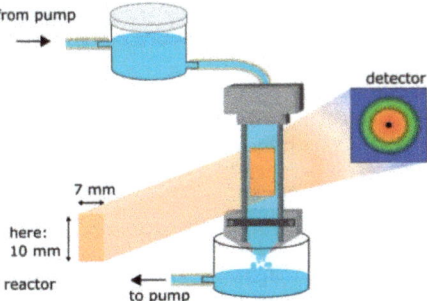

Figure 13. Container-less measuring device. A continuous stream of liquid is spread out by a nozzle between vertical wires. The free-standing film can be approximately matched to the size of a conventional neutron beam. Not shown is the bag filled with inert gas surrounding the system that is used to avoid the sample interacting with air moisture, resulting in an altered H-D ratio [121].

A rather specialized container-less measurement is the formation of soot particles in a combustion process. Here the flame is placed on a vertical translation stage so that the distance between the flame-neutron beam interception point and the flame mixer can be varied. This method has been used at a number of synchrotron and neutron facilities [122]. Just as in similar SAXS experiments, the data quality remains low and suitable caution is required when fitting the data.

Whatever the method used, one should be aware of the fact that container-less measurements do not mean that there is no sample-environment interaction. Oxidation, denaturing of protein, and exchange of deuterated for protiated water have been reported. It is also well-known that elongated molecules can orient themselves around air–water interfaces.

3.9. Ultrasound

Ultrasound used in aqueous solutions, or other solvents, creates cavitation bubbles that cause local agitation, affecting a variety of processes. Depending on the frequency used and physical system under investigation, ultrasound can be destructive. For example, it is used extensively for cleaning objects (20–40 kHz), but it can also help materials to crystallize (20 kHz) [123], or used as a diagnostic tool. Another application of ultrasound at a frequency around 1 MHz is to create a time-dependent mechanical perturbation. This was demonstrated in SANS experiments on sodium dodecyl sulfate (SDS) surfactant micelles in aqueous solution. Here the application of ultrasound allowed for the reversible observation of deformation of micelles as a function of time exposure to ultrasound [91].

A more elaborate ultrasound system that operates in the 1.25 MHz range, uses two spherically focused acoustic transducers and an acoustic cavitation detection system. The latter can be used to decouple changes observed due to cavitation versus changes that occur as a result of the propagation of non-cavitating acoustic waves [124]. In these experiments, a neutron absorbing aperture was used in order to match the dimensions of the neutron beam to those of the acoustic field. Even though the size of acoustic bubbles was within the detection range of the SANS experiments, it was found that due to their short half-lives they did not affect the data analysis to any appreciable extent. Several suggestions for future developments, beyond those in this manuscript, are given by the authors.

3.10. Humidity Control

In a range of fields, the control of humidity is important. For biological materials that have to be studied in their natural state, this is an obvious requirement, but industrial processes and the filling of the pores of porous materials to influence the scattering contrast should also be mentioned.

The simplest method is to place the samples in a closed environment with a reservoir containing a saturated salt solution. However, to change the relative humidity one has to use different salt solutions, which makes this method somewhat cumbersome for on-line experiments [125].

An elaborate system using mass flow controllers with the possibility to mix two different vapors to control the relative humidity between 0 and 90% has been used to study semicrystalline polymers, porous materials, and polyelectrolyte membranes [126]. The advantage of having two independent streams of vapor that can be mixed is that it offers the possibility to vary the H_2O/D_2O ratio, which in the case of porous materials, allows one to find the matching point for contrast variations very rapidly. The same system can be used to create a vapor pressure of organic solvents.

Developments in commercial humidity generators allow for the construction of rather simplified humidity-controlled cells, which were used in studies of forest products [127] where structural changes as function of moisture were observed. A commercial humidity generator was also used to study novel phases of lipid bilayers that are important for membrane fusion [128].

Relative humidity control of membranes in fuel cells during SANS experimentation was reported. However, in order to gain insights into the performance of nafion membranes in fuel cells under normal operating conditions, a total immersion of the membranes in H_2O/D_2O, whilst the material was uniaxially deformed, was required. To avoid a complicated experimental set-up with multiple mechanical feedthroughs to the liquid reservoir, U-shaped load beams were used, allowing for the material of interest to be immersed in a temperature-controlled D_2O reservoir [129]. See Figure 14.

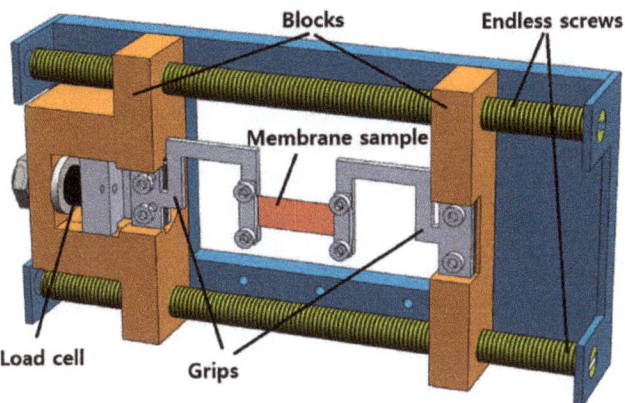

Figure 14. A load frame with U-shaped load beams that allow the sample to be deformed, whilst being completely immersed in water. The temperature-controlled reservoir containing H_2O/D_2O can be placed around the sample (reproduced with permission from [129], AIP Publishing, 2013).

3.11. Devices

Given the high penetration power of neutrons in combination with the possibility of highlighting different parts of a material or the distribution of water via the use of deuteration, one would guess that neutron scattering techniques would be a prime tool for the investigation of devices, like batteries, liquid/gas separating membranes under operando conditions, and soot formation by combustion of hydrocarbons in engines, to name a few. However, this kind of research does not currently have a widespread following. This might be due to a combination of the relatively low neutron fluxes available currently and a certain bias that neutron facilities are somewhat inaccessible to the general user. However, there are examples of the above-mentioned systems studied with neutrons. There also is somewhat of a bias against more engineering-oriented or applied applications that can be found in many beam line access panels. Most of the experiments using simulated devices can be found in energy storage and generation-related materials.

The device that was studied using neutron scattering and imaging techniques was a polymer-electrolyte fuel cell [130]. The interest here is the distribution and transport in the fuel cell and the two techniques, SANS and imaging, were combined in a single experiment where the neutron imaging system could be moved out of the beam during the acquisition of the SANS data. This is particularly interesting since the two techniques might not give the same type of information, but the accessible length scales are complementary, and relevant information over length scales from nano- to milli-meters can be obtained in a single experiment. It can be pointed out that this cell was a realistic model where metal components were illuminated by the neutron beam. Using SAXS experiments, where the electron densities of H_2O and polymers are not sufficiently different to generate a contrast that can make aluminum, H_2O and polymer all visible, no information about the distribution of water can be obtained. Importantly, the above-mentioned experiment emphasizes the ability to study materials under operando conditions. It is noteworthy that data acquisition time for SANS was 80 s/frame and for imaging 210 s/frame. In a later publication, infrared spectroscopic data were also used to evaluate the physical and chemical events inside the fuel cell [131,132].

Another energy-related application is the study of Li-ion batteries. The concentration of Li-salt and the type of electrode material used has consequences for the operation of the battery and its longevity. By using a half-cell, i.e., a single electrode surrounded by an electrolyte solution, it enables one to follow the events at the solid–electrolyte interface. In the case where one uses an ordered mesoporous carbon electrode, mesoporous events take place in the range where small angle scattering (SAXS/SANS) can yield information. Here again there is a lack of electron density contrast between the Carbon and the Li compared to the pores, which indicates that neutron scattering is better suited to gain insights into the effects of the use of different Li-salts and concentrations during the duty cycle for in-operando cycling/discharging [133]. Similar experiments, but with greater emphasis on the analysis of the scattering results and the role of the scattering contrast in the analysis of the data, were also published [134].

Supercapacitors, hybrid devices with characteristics between a capacitor and a battery, rely on a high internal surface for redox reactions to take place. The most promising materials are porous nitrides of vanadium and molybdenum, which strictly speaking, do not classify as soft matter, but in the context of this review are relevant as they are materials known to have some of the highest storage capacities [135]. By performing the required electrochemistry on-line, it was shown that the smallest pores in these materials allow for the increased adsorption of OH^- ions. These experiments would also benefit from on-line X-ray spectroscopy experiments in order to obtain insights about what is driving the charge storage mechanism.

4. In-Situ Technique Combinations

In most research, neutron scattering is only one of the tools in the experimental toolbox required to obtain the knowledge of the material properties that one is investigating. In certain circumstances, it is beneficial if one can combine neutron scattering experiments with complementary on-line auxiliary techniques, as has become a fairly routine approach at synchrotron radiation facilities [136,137]. The advantages of being able to simultaneously collect complementary data sets at the same time, using the same sample, outweigh any disadvantages due to the increased experimental complications regarding synchronization of the data sets and access of different probe beams, which cannot always use the same window material. Conditions in which one can consider the use of a multimodal approach are when dealing with spatially inhomogeneous materials or when the time resolution in an experiment is too fast to be able to stitch the data sets in a reliable way that does not create uncertainties in the time-correlation between the different experiments. In both cases, one should try to interrogate the same sample volume with the different probes.

The use of on-line Differential Scanning Calorimetry (DSC) with X-ray measurements is a good example where the data quality of the DSC signals suffers somewhat but where the synergy of having two simultaneous data sets outweighs the loss of quality [22]. When the strong peaks in a DSC curve can be correlated to the X-ray frame number, one can correlate the real thermodynamic temperature of the sample with the appropriate X-ray structure. Hence, the correlation can be made with an accurate DSC curve obtained with a well calibrated off-line instrument. The design problems for constructing a similar set-up for neutrons are somewhat larger than those for X-rays, since the beam access window needs to be considerably larger. However, this engineering issue can be overcome as was shown in phase separation experiments on partially deuterated alkanes, $C_nH2_{n+2}{:}C_mH(D)_{2n+2}$. A commercial instrument with an operation range from $-150\,°C$ to $+500\,°C$ was successfully modified to enable neutron beam access. A slight drawback is that it was not feasible to use commercially available DSC pans but instead special pans had to be custom designed and machined [138]. The time resolution that could be achieved in this experiment was 2 min/frame.

The combination of Raman scattering with different neutron scattering techniques has been reported over the years [139,140], including for the protein lysozyme, but these experiments were all carried out at cryogenic temperatures, which is of limited use in soft matter research. However, it was only recently that the cold deformation, i.e., ambient temperature, of low-density polyethylene was followed by both SANS and polarized Raman spectroscopy [141], where the experiments were performed using the same tensile stage but not simultaneously. By combining the different data sets, one can investigate the interplay between chain stretching (SANS) and the transition from amorphous to all trans conformers (Raman). Although these experiments were done separately, with the present generation of portable Raman spectrometers, there is no reason why this could not be carried out simultaneously if the required time-resolution would make this beneficial [142].

Dielectric spectroscopy can provide information at the molecular level about a variety of soft matter processes. For instance, chemical changes, segmental relaxation in polymers, and phase transitions are among the physical phenomena where dielectric spectroscopy can render complementary information to X-ray [143] and neutron scattering [144]. Such experiments were carried out in combination with a strain controlled Couette shear cell on a SANS instrument [145]. The Couette cylinders were constructed of titanium with thin mylar windows, allowing access to the neutron beam as well as for impedance measurements between the inner and outer cylinders (See Figure 15 for a schematic layout of the experiments). Crucial for this type of experiment is the synchronization of the different techniques, which can be readily achieved using 'event-mode' data collection [146]. Data were successfully obtained on the self-assembly of conjugated polymer melts and shear sensitive ionic liquids.

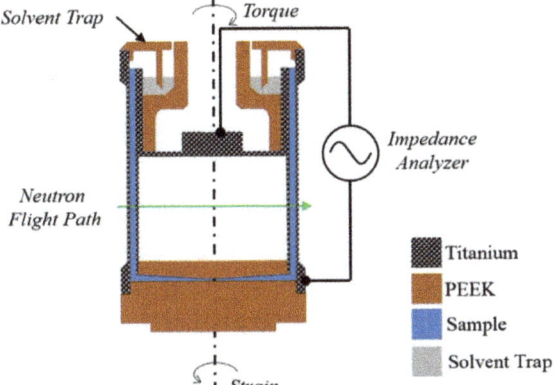

Figure 15. Schematic of a Couette rheology cell that allows for the simultaneous collection of rheological, dielectric spectroscopy, and neutron scattering data. The electrical impedance is measured between the Ti cylinders. Temperature control is achieved by using a forced convection oven that surrounds the cell (reproduced with permission from [145], AIP Publishing, 2017).

In order to investigate how high a molecular weight of polyethylene glycol dimethyl ether (PEGDME) can be incorporated in the crystalline region of syndiotactic polystyrene, one requires both the structural information provided by neutron scattering and vibrational spectroscopy. This was done by combining Fourier Transform Infrared Spectroscopy (FTIR) with SANS [147,148]. Here one encounters the issue that the optimum sample thickness required by both techniques is not the same. This can be partially overcome by using an FTIR sample cell and Attenuated Total Reflection, where the infrared beam makes multiple passes through the sample material [149]. However, a better solution is to place the IR and neutron beams co-parallel to ensure that one is studying the same part of the sample. This can be achieved by coupling the IR beam with an elaborate system of KBr crystals that act as mirrors for the IR beam but are relatively transparent to neutrons. As a compromise, the sample thickness was chosen to be 50 µm, just sufficiently thick for neutron scattering but thin enough for infra-red measurements. The combined data sets showed a distinct difference between the PEGDME conformation in the amorphous and crystalline regions of the polystyrene.

In experiments where the materials exhibit (partial) orientation and are transparent to visible light, birefringence is widely used to determine orientational effects. Neutron and X-ray scattering are also sensitive to orientational order, albeit on different length scales than birefringence. Scattering is mainly sensitive to the alignment of molecules, whilst birefringence is sensitive to bulk material properties [150]. Although such a capability seems to be very useful, it is somewhat surprising that this has not been applied more often [62,150,151].

Even though it is understood that having combined neutron and X-ray scattering data is highly beneficial in understanding a system, these techniques are not combined as often as they should [152–154]. In an ideal world, one would be able to use parallel beams of X-rays and neutrons, but an orthogonal combination is technically easier to realize. The achievable time-resolution, i.e., the length per time frame in which useable data can be collected, for the two experiments should be of the same order of magnitude. Until recently, such a combination was not feasible, but developments in X-ray generation and detector technology have allowed the first steps in allowing such a combined scattering instrument to be developed. Specifically, the two probe beams were placed at right angles to one another, making the system less useful for samples that are anisotropic, but it can serve as a proof of principle [155]. The complication that one faces with the development of such an instrument is that one invariably tries to mount an X-ray system on an existing neutron scattering beamline. Hence, one encounters space and shielding issues that make things more complicated (but not impossible).

The combination of neutron scattering with light scattering or birefringence should be more straightforward to achieve and would extend the length scales over which information can be obtained. However, this also seems to be an area waiting to be developed since few publications can be found. A combination of SANS with diffuse wave spectroscopy (DWS) was reported. DWS can be used in high concentration colloid solutions and can be used to obtain information about the ensemble dynamics of particles, while SANS provides the structural information [156].

An example where an industrially relevant testing method, Rapid Visco Analysis (RVA), was combined with SANS experiments was used to determine the nature of the structural changes in starch during pasting. This combined mechanical and thermal treatment of materials plays an important role in process- and quality-control in the food industries. By combining the neutron scattering data with the viscosity, as measured with an adapted RVA system, the abrupt changes in viscosity could be correlated with the changing morphology of the starch and the starch derivative network [157].

5. New Directions

One of the disadvantages of neutron scattering experiments is the size of the neutron beam. Due to the abundance of available photons and the development of optical systems that take advantage of this abundance, X-ray beam sizes on the submicron scale are common. This is not the case with neutrons. A smaller neutron beam is achieved at the expense of available flux, and hence of the attainable time resolution. Although one cannot reasonably expect sub-micron neutron beams, it is still feasible to obtain spatially resolved data where the (inhomogeneous) sample is probed with a small beam of around 100 microns at different positions [48] or to use microfluidic devices [157].

The promise of a higher effective flux that will be made available by the new generation of spallation sources will obviously have an impact on how fast data can be collected. However, this increase in intensity will not bring many orders of magnitude of improvement. Therefore, if one wants to increase the potential time-resolution, one should also look at better ways to perform experiments and at innovative ways of processing the data, and where the emphasis is placed on which parameters one would like to measure instead of trying to obtain the best statistical quality data. Certainly, for non-isotropic scattering samples there are improvements possible in what kind of time-resolution can be achieved.

The combination of neutron scattering, or spectroscopy data and complementary techniques is quite common. Less common, however, are approaches to analyze the data simultaneously, for example, using multidimensional correlation spectroscopy [158,159]. This method has already been applied in X-ray scattering experiments to elucidate events at the early onset of crystallization, where the nascent signals of the new phases are very weak [160,161], and the impact of radiation damage on proteins and the assembly in nanomaterials [162], and can be considered reminiscent of the low statistical data quality one would encounter in fast time-resolved neutron experiments.

So far, we have not mentioned the use of computational modelling, but it is not unreasonable to assume that this can in the near future become a virtual on-line technique combination. Modeling can become an inroad for new approaches to neutron experiments, which are optimized to provide just sufficient data quality to resolve whether the modeling result can be confirmed or disproven. This would require a 'pipeline' that feeds from modeling to defining neutron experiments.

An interesting development is the advances in ray-tracing simulations for instrument designs, e.g., McStas [163] or McVine [164]. These programs have expanded to include sample scattering kernels and sample environments [165]. It is now possible to simulate accurate sample geometries, instrument resolution, and counting statistics [166,167] to determine if planned experiments are feasible.

The importance of thin polymer films and grazing incidence studies characterizing the structure and structure formation of materials has increased in recent years [168]. In the case of X-ray studies, complicated on-line experiments, like roll-to-roll printing of polymers, have been reported. Neutron-based grazing incidence studies undoubtedly can play a bigger role in elucidating more complex morphologies. Data collection times, however, are still on the order of hours for a single static measurement [169], although in favorable cases, this can be reduced to minutes and can be further improved when dealing with cyclic processes [18]. It should also be noted that these results were obtained on non-dedicated beamlines. With a specialized design, further improvements could be made.

With the higher flux that one can expect from the new neutron sources that are being developed, obviously the situation for 'neutron hungry techniques', like time-resolved, spatially-resolved, and grazing incidence scattering experiments will improve. However, one cannot expect the many decades in improvement, which was the signature of the successive generations of X-ray sources, to be achievable. Still, there is substantial scope for increasing the experimental sophistication in soft matter research using neutron scattering techniques.

6. Conclusions

This article gives an overview of diverse sample environments that can be used in neutron scattering techniques applied in the area of soft condensed matter. We have focused on the small angle neutron scattering technique. Even though we have covered many different areas of sample environment control, such a survey can never be all-encompassing. However, we hope to have demonstrated with this overview that neutron scattering is not only suitable for static structure determination but can be implemented in a greater number of studies than what is being used presently. We suggest that more emphasis should be placed in SANS applications on observing the time-evolution of selected structural parameters during the response of materials to external perturbations. In this way, neutron scattering has the potential to make impactful contributions to the understanding of complex systems, such as soft materials, even when a complete structural analysis is out of reach. Importantly, observing parametrized changes in materials often does not require the highest statistical data quality, and therefore improvements in time resolution can be found by carefully optimizing to the minimum counting statistics level that is necessary, rather than following traditional optimization rules that were developed for obtaining best data from static structures.

Author Contributions: W.B. carried out the literature investigation, all authors contributed equally to the manuscript writing. All authors have read and agreed to the published version of the manuscript.

Funding: This research received no external funding.

Acknowledgments: We would like to acknowledge input from our colleagues Ralf Schweins, Steve King, Paul Butler, Sai Venkatesh Pingali, Frank Bates, Bin Wu and Gary Lynn. A special acknowledgement is due for our sadly missed colleague Isabelle Grillo. T. Plivelic, J. Vermant and R. Winter have made original figures available. The anonymous reviewers have made some valuable suggestions and are acknowledged for their careful reading of the manuscript. W.B.'s contribution is based upon work supported by Oak Ridge National Laboratory, managed by UT-Battelle LLC, for the U.S. Department of Energy. A portion of this research used resources at the High Flux Isotope Reactor and the Spallation Neutron Source, DOE Office of Science User Facilities operated by the Oak Ridge National Laboratory. DOE will provide public access to these results of federally sponsored research in accordance with the DOE Public Access Plan http://energy.gov/downloads/doe-public-access-plan (accessed on 13 May 2021).

Conflicts of Interest: The authors declare no conflict of interest.

References

1. Hemberg, O.; Otendal, M.; Hertz, H.M. Liquid-metal-jet anode X-ray tube. *Opt. Eng.* **2004**, *43*, 1682–1688. [CrossRef]
2. Bauters, S.; Tack, P.; Rudloff-Grund, J.H.; Banerjee, D.; Longo, A.; Vekemans, B.; Bras, W.; Brenker, F.E.; van Silfhout, R.; Vincze, L. Polycapillary Optics Based Confocal Micro X-ray Fluorescence and X-ray Absorption Spectroscopy Setup at The European Synchrotron Radiation Facility Collaborative Research Group Dutch-Belgian Beamline. *Anal. Chem.* **2018**, *90*, 2389–2394. [CrossRef] [PubMed]
3. Donath, T.; Brandstetter, S.; Cibik, L.; Commichau, S.; Hofer, P.; Krumrey, M.; Luthi, B.; Marggraf, S.; Muller, P.; Schneebeli, M.; et al. Characterization of the PILATUS photon-counting pixel detector for X-ray energies from 1.75 keV to 60 keV. In Proceedings of the 11th International Conference on Synchrotron Radiation Instrumentation (Sri 2012), Lyon, France, 9–13 July 2012; p. 425.
4. Jinnai, H.; Hasegawa, H.; Hashimoto, T.; Han, C.C. Time-resolved small-angle neutron-scattering in intermediate-stage and late-stage spinodal decomposition of dpb hpi blends. *Macromolecules* **1991**, *24*, 282–289. [CrossRef]
5. Connell, J.G.; Richards, R.W.; Rennie, A.R. Phase-separation kinetics in concentrated-solutions of linear diblock copolymers of polystyrene and polyisoprene from time-resolved small-angle neutron-scattering. *Polymer* **1991**, *32*, 2033–2042. [CrossRef]
6. Egelhaaf, S.U.; Schurtenberger, P. Micelle-to-vesicle transition: A time-resolved structural study. *Phys. Rev. Lett.* **1999**, *82*, 2804–2807. [CrossRef]
7. Hollamby, M.J.; Borisova, D.; Brown, P.; Eastoe, J.; Grillo, I.; Shchukin, D. Growth of Mesoporous Silica Nanoparticles Monitored by Time-Resolved Small-Angle Neutron Scattering. *Langmuir* **2012**, *28*, 4425–4433. [CrossRef]
8. Niimura, N.; Minezaki, Y.; Ataka, M.; Katsura, T. Aggregation in supersaturated lysozyme solution studied by time-resolved small-angle neutron-scattering. *J. Cryst. Growth* **1995**, *154*, 136–144. [CrossRef]
9. Mahabir, S.; Small, D.; Li, M.; Wan, W.K.; Kucerka, N.; Littrell, K.; Katsaras, J.; Nieh, M.P. Growth kinetics of lipid-based nanodiscs to unilamellar vesicles-A time-resolved small angle neutron scattering (SANS) study. *Biochim. Biophys. Acta-Biomembr.* **2013**, *1828*, 1025–1035. [CrossRef]
10. Mason, T.G.; Lin, M.Y. Time-resolved small angle neutron scattering measurements of asphaltene nanoparticle aggregation kinetics in incompatible crude oil mixtures. *J. Chem. Phys.* **2003**, *119*, 565–571. [CrossRef]
11. Lopez-Barron, C.R.; Wagner, N.J.; Porcar, L. Layering, melting, and recrystallization of a close-packed micellar crystal under steady and large-amplitude oscillatory shear flows. *J. Rheol.* **2015**, *59*, 793–820. [CrossRef]
12. Lopez-Barron, C.R.; Gurnon, A.K.; Eberle, A.P.R.; Porcar, L.; Wagner, N.J. Microstructural evolution of a model, shear-banding micellar solution during shear startup and cessation. *Phys. Rev. E* **2014**, *89*, 11. [CrossRef]
13. Wrede, O.; Reimann, Y.; Lulsdorf, S.; Emmrich, D.; Schneider, K.; Schmid, A.J.; Zauser, D.; Hannappel, Y.; Beyer, A.; Schweins, R.; et al. Volume phase transition kinetics of smart N-n-propylacrylamide microgels studied by time-resolved pressure jump small angle neutron scattering. *Sci. Rep.* **2018**, *8*, 13. [CrossRef]
14. Terrill, N.J.; Fairclough, P.A.; Towns-Andrews, E.; Komanschek, B.U.; Young, R.J.; Ryan, A.J. Density fluctuations: The nucleation event in isotactic polypropylene crystallization. *Polymer* **1998**, *39*, 2381–2385. [CrossRef]
15. Metze, M.; Barbe, S.; Reiche, A.; Kesting, A.; Schweins, R. A Neutron-Transparent Flow-Through Cell (NTFT-Cell) for the SANS investigation of microstructure evolution during industrial evaporative casting. *J. Neutron Res.* **2017**, *19*, 177–185. [CrossRef]
16. Kipping, D.; Gahler, R.; Habicht, K. Small angle neutron scattering at very high time resolution: Principle and simulations of 'TISANE'. *Phys. Lett. A* **2008**, *372*, 1541–1546. [CrossRef]

17. Wiedenmann, A.; Keiderling, U.; Habicht, K.; Russina, M.; Gahler, R. Dynamics of field-induced ordering in magnetic colloids studied by new time-resolved small-angle neutron-scattering techniques. *Phys. Rev. Lett.* **2006**, *97*, 4. [CrossRef] [PubMed]
18. Adlmann, F.A.; Gutfreund, P.; Ankner, J.F.; Browning, J.F.; Parizzi, A.; Vacaliuc, B.; Halbert, C.E.; Rich, J.P.; Dennison, A.J.C.; Wolff, M.R. Towards neutron scattering experiments with sub-millisecond time resolution. *J. Appl. Crystallogr.* **2015**, *48*, 220–226. [CrossRef]
19. Levelut, A.M.; Guinier, A. X-ray Scattering by Point Defects. In *Small Angle X-ray Scattering*; Brumberger, H., Ed.; Gordon and Breach: New York, NY, USA, 1967.
20. Olsson, A.; Hellsing, M.S.; Rennie, A.R. A holder to rotate sample cells to avoid sedimentation in small-angle neutron scattering and ultra small-angle neutron scattering experiments. *Meas. Sci. Technol.* **2013**, *24*, 6. [CrossRef]
21. Leao, J.B.; Murphy, R.P.; Wagner, N.J.; Bleuel, M. Dynamic infrared sample controlled (DISCO) temperature for the tumbler cells for ultra small angle neutron scattering (USANS). *J. Neutron Res.* **2017**, *19*, 23–26. [CrossRef]
22. Bras, W.; Derbyshire, G.E.; Devine, A.; Clark, S.M.; Cooke, J.; Komanschek, B.E.; Ryan, A.J. The Combination of Thermal-Analysis and Time-Resolved X-Ray Techniques—A Powerful Method for Materials Characterization. *J. Appl. Crystallogr.* **1995**, *28*, 26–32. [CrossRef]
23. Russell, T.P.; Koberstein, J.T. Simultaneous Differential Scanning Calorimetry And Small-Angle X-Ray-Scattering. *J. Polym. Sci. Part B-Polym. Phys.* **1985**, *23*, 1109–1115. [CrossRef]
24. Gilbert, B. Finite size effects on the real-space pair distribution function of nanoparticles. *J. Appl. Crystallogr.* **2008**, *41*, 554–562. [CrossRef]
25. Oskolkova, M.Z.; Norrman, E.; Olsson, U. Study of the micelle-to-vesicle transition and smallest possible vesicle size by temperature-jumps. *J. Colloid Interface Sci.* **2013**, *396*, 173–177. [CrossRef] [PubMed]
26. Adelsberger, J.; Grillo, I.; Kulkarni, A.; Sharp, M.; Bivigou-Koumba, A.M.; Laschewsky, A.; Müller-Buschbaum, P.; Papadakis, C.M. Kinetics of aggregation in micellar solutions of thermoresponsive triblock copolymers—Influence of concentration, start and target temperatures. *Soft Matter* **2013**, *9*, 1685–1699. [CrossRef]
27. Pullen, S.A.; Gilbert, E.P.; Olsen, S.R.; Lang, E.A.; Doolan, K.R. An in situ rapid heat-quench cell for small-angle neutron scattering. *Meas. Sci. Technol.* **2008**, *19*, 8. [CrossRef]
28. Adelsberger, J.; Metwalli, E.; Diethert, A.; Grillo, I.; Bivigou-Koumba, A.M.; Laschewsky, A.; Muller-Buschbaum, P.; Papadakis, C.M. Kinetics of Collapse Transition and Cluster Formation in a Thermoresponsive Micellar Solution of P(S-b-NIPAM-b-S) Induced by a Temperature Jump. *Macromol. Rapid Commun.* **2012**, *33*, 254–259. [CrossRef]
29. Dewers, T.A.; Heath, J.E.; Bryan, C.R.; Mang, J.T.; Hjelm, R.P.; Ding, M.; Taylor, M. Oedometric Small-Angle Neutron Scattering: In Situ Observation of Nanopore Structure During Bentonite Consolidation and Swelling in Dry and Hydrous CO_2 Environments. *Environ. Sci. Technol.* **2018**, *52*, 3758–3768. [CrossRef]
30. Stefanopoulos, K.L.; Youngs, T.G.A.; Sakurovs, R.; Ruppert, L.F.; Bahadur, J.; Melnichenko, Y.B. Neutron Scattering Measurements of Carbon Dioxide Adsorption in Pores within the Marcellus Shale: Implications for Sequestration. *Environ. Sci. Technol.* **2017**, *51*, 6515–6521. [CrossRef]
31. Hjelm, R.P.; Taylor, M.A.; Frash, L.P.; Hawley, M.E.; Ding, M.; Xu, H.W.; Barker, J.; Olds, D.; Heath, J.; Dewers, T. Flow-through compression cell for small-angle and ultra-small-angle neutron scattering measurements. *Rev. Sci. Instrum.* **2018**, *89*, 9. [CrossRef]
32. Gabke, A.; Kraineva, J.; Kohling, R.; Winter, R. Using pressure in combination with x-ray and neutron scattering techniques for studying the structure, stability and phase behaviour of soft condensed matter and biomolecular systems. *J. Phys. Condens. Matter* **2005**, *17*, S3077–S3092. [CrossRef]
33. Teixeira, S.C.M.; Leao, J.B.; Gagnon, C.; McHugh, M.A. High pressure cell for Bio-SANS studies under sub-zero temperatures or heat denaturing conditions. *J. Neutron Res.* **2018**, *20*, 11–21. [CrossRef]
34. Pingali, S.V.; Smith, M.D.; Liu, S.-H.; Rawal, T.B.; Pu, Y.; Shah, R.; Evans, B.R.; Urban, V.S.; Davison, B.H.; Cai, C.M.; et al. Deconstruction of biomass enabled by local demixing of cosolvents at cellulose and lignin surfaces. *Proc. Natl. Acad. Sci. USA* **2020**, *117*, 16776–16781. [CrossRef]
35. Niebuur, B.-J.; Chiappisi, L.; Zhang, X.; Jung, F.; Schulte, A.; Papadakis, C.M. Formation and Growth of Mesoglobules in Aqueous Poly(N-isopropylacrylamide) Solutions Revealed with Kinetic Small-Angle Neutron Scattering and Fast Pressure Jumps. *ACS Macro Lett.* **2018**, *7*, 1155–1160. [CrossRef]
36. Ryan, A.J.; Bras, W.; Hermida-Merino, D.; Cavallo, D. The interaction between fundamental and industrial research and experimental developments in the field of polymer crystallization. *J. Non-Cryst. Solids* **2016**, *451*, 168–178. [CrossRef]
37. Roozemond, P.C.; Ma, Z.; Cui, K.; Li, L.; Peters, G.W.M. Multimorphological Crystallization of Shish-Kebab Structures in Isotactic Polypropylene: Quantitative Modeling of Parent-Daughter Crystallization Kinetics. *Macromolecules* **2014**, *47*, 5152–5162. [CrossRef]
38. Lindner, P.; Oberthur, R.C. Shear induced deformation of polystyrene coils in dilute-solution from small-angle neutron-scattering. 1. Shear gradient apparatus and 1st experimental results. *Colloid Polym. Sci.* **1985**, *263*, 443–453. [CrossRef]
39. Eberle, A.P.R.; Porcar, L. Flow-SANS and Rheo-SANS applied to soft matter. *Curr. Opin. Colloid Interface Sci.* **2012**, *17*, 33–43. [CrossRef]
40. Huang, G.R.; Wang, Y.Y.; Wu, B.; Wang, Z.; Do, C.; Smith, G.S.; Bras, W.; Porcar, L.; Falus, P.; Chen, W.R. Reconstruction of three-dimensional anisotropic structure from small-angle scattering experiments. *Phys. Rev. E* **2017**, *96*, 022612. [CrossRef] [PubMed]

41. Porcar, L.; Pozzo, D.; Langenbucher, G.; Moyer, J.; Butler, P.D. Rheo-small-angle neutron scattering at the National Institute of Standards and Technology Center for Neutron Research. *Rev. Sci. Instrum.* **2011**, *82*, 7. [CrossRef]
42. Velichko, E.; Tian, B.; Nikolaeva, T.; Koning, J.; van Duynhoven, J.; Bouwman, W.G. A versatile shear cell for investigation of structure of food materials under shear. *Colloids Surf. A-Physicochem. Eng. Asp.* **2019**, *566*, 21–28. [CrossRef]
43. Vermant, J.; Solomon, M.J. Flow-induced structure in colloidal suspensions. *J. Phys. Condens. Matter* **2005**, *17*, R187–R216. [CrossRef]
44. Helgeson, M.E.; Vasquez, P.A.; Kaler, E.W.; Wagner, N.J. Rheology and spatially resolved structure of cetyltrimethylammonium bromide wormlike micelles through the shear banding transition. *J. Rheol.* **2009**, *53*, 727–756. [CrossRef]
45. Nusser, K.; Neueder, S.; Schneider, G.J.; Meyer, M.; Pyckhout-Hintzen, W.; Willner, L.; Radulescu, A.; Richter, D. Conformations of silica− poly (ethylene− propylene) nanocomposites. *Macromolecules* **2010**, *43*, 9837–9847. [CrossRef]
46. Gasnier, A.; Royal, G.; Terech, P. Metallo-Supramolecular Gels Based on a Multitopic Cyclam Bis-Terpyridine Platform. *Langmuir* **2009**, *25*, 8751–8762. [CrossRef] [PubMed]
47. Wang, Z.; Iwashita, T.; Porcar, L.; Wang, Y.Y.; Liu, Y.; Sanchez-Diaz, L.E.; Wu, B.; Huang, G.R.; Egami, T.; Chen, W.R. Local elasticity in nonlinear rheology of interacting colloidal glasses revealed by neutron scattering and rheometry. *Phys. Chem. Chem. Phys.* **2019**, *21*, 38–45. [CrossRef]
48. Liberatore, M.W.; Nettesheim, F.; Wagner, N.J.; Porcar, L. Spatially resolved small-angle neutron scattering in the 1-2 plane: A study of shear-induced phase-separating wormlike micelles. *Phys. Rev. E* **2006**, *73*, 020504. [CrossRef]
49. Kim, J.; Helgeson, M.E. Shear-induced clustering of Brownian colloids in associative polymer networks at moderate Peclet number. *Phys. Rev. Fluids* **2016**, *1*, 19. [CrossRef]
50. Sharma, J.; King, S.M.; Bohlin, L.; Clarke, N. Apparatus for simultaneous rheology and small-angle neutron scattering from high-viscosity polymer melts and blends. *Nucl. Instrum. Methods Phys. Res. Sect. A-Accel. Spectrometers Detect. Assoc. Equip.* **2010**, *620*, 437–444. [CrossRef]
51. Rexeisen, E.L.; Fan, W.; Pangburn, T.O.; Taribagil, R.R.; Bates, F.S.; Lodge, T.P.; Tsapatsis, M.; Kokkoli, E. Self-Assembly of Fibronectin Mimetic Peptide-Amphiphile Nanofibers. *Langmuir* **2010**, *26*, 1953–1959. [CrossRef] [PubMed]
52. Sasa, L.A.; Yearley, E.J.; Welch, C.F.; Taylor, M.A.; Gilbertson, R.D.; Hammeter, C.; Majewski, J.; Hjelm, R.P. The Los Alamos Neutron Science Center neutron rheometer in the cone and plate geometry to examine tethered polymers/polymer melt interfaces via neutron reflectivity. *Rev. Sci. Instrum.* **2010**, *81*, 6. [CrossRef]
53. Corvis, Y.; Barre, L.; Jestin, J.; Gummel, J.; Cousin, F. Asphaltene adsorption mechanism under shear flow probed by in situ neutron reflectivity measurements. *Eur. Phys. J. Spec. Top.* **2012**, *213*, 295–302. [CrossRef]
54. Hamley, I.W.; Burholt, S.; Hutchinson, J.; Castelletto, V.; da Silva, E.R.; Alves, W.; Gutfreund, P.; Porcar, L.; Dattani, R.; Hermida-Merino, D.; et al. Shear Alignment of Bola-Amphiphilic Arginine-Coated Peptide Nanotubes. *Biomacromolecules* **2017**, *18*, 141–149. [CrossRef] [PubMed]
55. Muller, R.; Picot, C.; Zang, Y.H.; Froelich, D. Polymer-chain conformation in the melt during steady elongational flow as measured by sans—Temporary network model. *Macromolecules* **1990**, *23*, 2577–2582. [CrossRef]
56. Ruocco, N.; Auhl, D.; Bailly, C.; Lindner, P.; Pyckhout-Hintzen, W.; Wischnewski, A.; Leal, L.G.; Hadjichristidis, N.; Richter, D. Branch Point Withdrawal in Elongational Startup Flow by Time-Resolved Small Angle Neutron Scattering. *Macromolecules* **2016**, *49*, 4330–4339. [CrossRef]
57. Macosko, C.W.; Mewis, J. *Suspension Rheology*; Wiley-VCH: New York, NY, USA, 1994.
58. Camargo, R.; Macosko, C.; Tirrell, M.; Wellinghoff, S. Phase separation studies in RIM polyurethanes catalyst and hard segment crystallinity effects. *Polymer* **1985**, *26*, 1145–1154. [CrossRef]
59. Inkson, N.J.; McLeish, T.C.B.; Harlen, O.G.; Groves, D.J. Predicting low density polyethylene melt rheology in elongational and shear flows with "pom-pom" constitutive equations. *J. Rheol.* **1999**, *43*, 873–896. [CrossRef]
60. Bent, J.F.; Richards, R.W.; Gough, T.D. Recirculation cell for the small-angle neutron scattering investigation of polymer melts in flow. *Rev. Sci. Instrum.* **2003**, *74*, 4052–4057. [CrossRef]
61. Bent, J.; Hutchings, L.R.; Richards, R.W.; Gough, T.; Spares, R.; Coates, P.D.; Grillo, I.; Harlen, O.G.; Read, D.J.; Graham, R.S.; et al. Neutron-mapping polymer flow: Scattering, flow visualization, and molecular theory. *Science* **2003**, *301*, 1691–1695. [CrossRef]
62. Graham, R.S.; Bent, J.; Hutchings, L.R.; Richards, R.W.; Groves, D.J.; Embery, J.; Nicholson, T.M.; McLeish, T.C.B.; Likhtman, A.E.; Harlen, O.G.; et al. Measuring and predicting the dynamics of linear monodisperse entangled polymers in rapid flow through an abrupt contraction. A small angle neutron scattering study. *Macromolecules* **2006**, *39*, 2700–2709.
63. McLeish, T.C.B.; Clarke, N.; de Luca, E.; Hutchings, L.R.; Graham, R.S.; Gough, T.; Grillo, I.; Fernyhough, C.M.; Chambon, P. Neutron flow-mapping: Multiscale modelling opens a new experimental window. *Soft Matter* **2009**, *5*, 4426–4432. [CrossRef]
64. Lee, J.S.; Dylla-Spears, R.; Teclemariam, N.P.; Muller, S.J. Microfluidic four-roll mill for all flow types. *Appl. Phys. Lett.* **2007**, *90*, 3. [CrossRef]
65. Taylor, G.I. The formation of emulsions in definable fields of flow. *Proc. R. Soc. A* **1934**, *146*, 501–523.
66. Corona, P.T.; Ruocco, N.; Weigandt, K.M.; Leal, L.G.; Helgeson, M.E. Probing flow-induced nanostructure of complex fluids in arbitrary 2D flows using a fluidic four-roll mill (FFoRM). *Sci. Rep.* **2018**, *8*, 18. [CrossRef]
67. Straube, E.; Urban, V.; Pyckhouthintzen, W.; Richter, D. Sans investigations of topological constraints and microscopic deformations in rubberelastic networks. *Macromolecules* **1994**, *27*, 7681–7688. [CrossRef]

68. Butler, M.F.; Donald, A.M.; Bras, W.; Mant, G.R.; Derbyshire, G.E.; Ryan, A.J. A real-time simultaneous small-angle and wide-angle X-ray-scattering study of in-situ deformation of isotropic polyethylene. *Macromolecules* **1995**, *28*, 6383–6393. [CrossRef]
69. Coutry, S.; Spells, S.J. Molecular changes on drawing isotopic blends of polyethylene and ethylene copolymers: 1. Static and time-resolved sans studies. *Polymer* **2003**, *44*, 1949–1956. [CrossRef]
70. Wiyatno, W.; Fuller, G.G.; Pople, J.A.; Gast, A.P.; Chen, Z.R.; Waymouth, R.M.; Myers, C.L. Component stress-strain behavior and small-angle neutron scattering investigation of stereoblock elastomeric polypropylene. *Macromolecules* **2003**, *36*, 1178–1187. [CrossRef]
71. Engqvist, J.; Hall, S.A.; Wallin, M.; Ristinmaa, M.; Plivelic, T.S. Multi-scale Measurement of (Amorphous) Polymer Deformation: Simultaneous X-ray Scattering, Digital Image Correlation and In-situ Loading. *Exp. Mech.* **2014**, *54*, 1373–1383. [CrossRef]
72. Wagner, J.A.; Patil, S.P.; Greving, I.; Lammel, M.; Gkagkas, K.; Seydel, T.; Muller, M.; Markert, B.; Grater, F. Stress-induced long-range ordering in spider silk. *Sci. Rep.* **2017**, *7*, 6. [CrossRef]
73. Stribeck, N.; Nochel, U.; Funari, S.S.; Schubert, T. Tensile tests of polypropylene monitored by SAXS. Comparing the stretch-hold technique to the dynamic technique. *J. Polym. Sci. Part B-Polym. Phys.* **2008**, *46*, 721–726. [CrossRef]
74. Dokter, W.H.; Vangarderen, H.F.; Beelen, T.P.M.; Vansanten, R.A.; Bras, W. Homogeneous Versus Heterogeneous Zeolite Nucleation. *Angew. Chem. -Int. Ed. Engl.* **1995**, *34*, 73–75. [CrossRef]
75. Grillo, I. Applications of stopped-flow in SAXS and SANS. *Curr. Opin. Colloid Interface Sci.* **2009**, *14*, 402–408. [CrossRef]
76. Virtanen, O.L.J.; Kather, M.; Meyer-Kirschner, J.; Melle, A.; Radulescu, A.; Viell, J.; Mitsos, A.; Pich, A.; Richtering, W. Direct Monitoring of Microgel Formation during Precipitation Polymerization of N-Isopropylacrylamide Using in Situ SANS. *ACS Omega* **2019**, *4*, 3690–3699. [CrossRef]
77. Uhríková, D.; Teixeira, J.; Hubčík, L.; Búcsi, A.; Kondela, T.; Murugova, T.; Ivankov, O.I. Lipid based drug delivery systems: Kinetics by SANS. *J. Phys. Conf. Ser.* **2017**, *848*, 012007. [CrossRef]
78. Hayward, D.W.; Chiappisi, L.; Prevost, S.; Schweins, R.; Gradzielski, M. A Small-Angle Neutron Scattering Environment for In-Situ Observation of Chemical Processes. *Sci. Rep.* **2018**, *8*, 2–11. [CrossRef] [PubMed]
79. Melnichenko, Y.B.; He, L.L.; Sakurovs, R.; Kholodenko, A.L.; Blach, T.; Mastalerz, M.; Radlinski, A.P.; Cheng, G.; Mildner, D.F.R. Accessibility of pores in coal to methane and carbon dioxide. *Fuel* **2012**, *91*, 200–208. [CrossRef]
80. Woods, H.M.; Silva, M.; Nouvel, C.; Shakesheff, K.M.; Howdle, S.M. Materials processing in supercritical carbon dioxide: Surfactants, polymers and biomaterials. *J. Mater. Chem.* **2004**, *14*, 1663–1678. [CrossRef]
81. Melnichenko, Y.B.; Kiran, E.; Wignall, G.D.; Heath, K.D.; Salaniwal, S.; Cochran, H.D. Stamm, Pressure- and temperature-induced transitions in solutions of poly(dimethylsiloxane) in supercritical carbon dioxide. M. *Macromolecules* **1999**, *32*, 5344–5347. [CrossRef]
82. Muller, A.; Putz, Y.; Oberhoffer, R.; Becker, N.; Strey, R.; Wiedenmann, A.; Sottmann, T. Kinetics of pressure induced structural changes in super- or near-critical CO_2-microemulsions. *Phys. Chem. Chem. Phys.* **2014**, *16*, 18092–18097. [CrossRef] [PubMed]
83. Putz, Y.; Grassberger, L.; Lindner, P.; Schweins, R.; Strey, R.; Sottmann, T. Unexpected efficiency boosting in CO_2-microemulsions: A cyclohexane depletion zone near the fluorinated surfactants evidenced by a systematic SANS contrast variation study. *Phys. Chem. Chem. Phys.* **2015**, *17*, 6122–6134. [CrossRef] [PubMed]
84. Hermida-Merino, D.; Portale, G.; Fields, P.; Wilson, R.; Bassett, S.P.; Jennings, J.; Dellar, M.; Gommes, C.; Howdle, S.M.; Vrolijk, B.C. A high pressure cell for supercritical CO_2 on-line chemical reactions studied with x-ray techniques. *Rev. Sci. Instrum.* **2014**, *85*, 093905. [CrossRef] [PubMed]
85. Jordan, A.; Jacques, M.; Merrick, C.; Devos, J.; Forsyth, V.T.; Porcar, L.; Martel, A. SEC-SANS: Size exclusion chromatography combined in situ with small-angle neutron scattering. *J. Appl. Crystallogr.* **2016**, *49*, 2015–2020. [CrossRef]
86. Koch, M.H.J.; Sayers, Z.; Sicre, P.; Svergun, D. A synchrotron-radiation electric-field X-ray solution scattering study of dna at very-low ionic-strength. *Macromolecules* **1995**, *28*, 4904–4907. [CrossRef]
87. Xu, T.; Zhu, Y.; Gido, S.P.; Russell, T.P. Electric Field Alignment of Symmetric Diblock Copolymer Thin Films. *Macromolecules* **2004**, *37*, 2625–2629. [CrossRef]
88. Pester, C.; Ruppel, M.; Schoberth, H.G.; Schmidt, K.; Liedel, C.; van Rijn, P.; Schindler, K.A.; Hiltl, S.; Czubak, T.; Mays, J.; et al. Piezoelectric Properties of Non-Polar Block Copolymers. *Adv. Mater.* **2011**, *23*, 4047–4052. [CrossRef]
89. Boker, A.; Ruppel, M.; Lindner, P.; Urban, V.S.; Schmidt, K.; Schoberth, H. *SANS Study on the Phase Behaviour of Block Copolymers in the Presence of an External Electric Field*; Institut Laue-Langevin (ILL) ILL User Club: Grenoble, France, 2009.
90. McMullan, J.M.; Wagner, N.J. Directed self-assembly of colloidal crystals by dielectrophoretic ordering observed with small angle neutron scattering (SANS). *Soft Matter* **2010**, *6*, 5443–5450. [CrossRef]
91. Li, D.S.; Lee, Y.-T.; Xi, Y.; Pelivanov, I.; O'Donnell, M.; Pozzo, L.D. A small-angle scattering environment for in situ ultrasound studies. *Soft Matter* **2018**, *14*, 5283–5293. [CrossRef] [PubMed]
92. Hamley, I.W.; Castelletto, V.; Lu, Z.B.; Imrie, C.T.; Itoh, T.; Al-Hussein, M. Interplay between Smectic Ordering and Microphase Separation in a Series of Side-Group Liquid-Crystal Block Copolymers. *Macromolecules* **2004**, *37*, 4798–4807. [CrossRef]
93. Hocine, S.; Brulet, A.; Jia, L.; Yang, J.; Di Cicco, A.; Bouteiller, L.; Li, M.H. Structural changes in liquid crystal polymer vesicles induced by temperature variation and magnetic fields. *Soft Matter* **2011**, *7*, 2613–2623. [CrossRef]
94. Nieh, M.P.; Glinka, C.J.; Krueger, S.; Prosser, R.S.; Katsaras, J. SANS study of the structural phases of magnetically alignable lanthanide-doped phospholipid mixtures. *Langmuir* **2001**, *17*, 2629–2638. [CrossRef]

95. van der Beek, D.; Petukhov, A.V.; Davidson, P.; Ferre, J.; Jamet, J.P.; Wensink, H.H.; Vroege, G.J.; Bras, W.; Lekkerkerker, H.N.W. Magnetic-field-induced orientational order in the isotropic phase of hard colloidal platelets. *Phys. Rev. E* **2006**, *73*, 10. [CrossRef] [PubMed]
96. McCulloch, B.; Portale, G.; Bras, W.; Segalman, R.A. Increased Order-Disorder Transition Temperature for a Rod-Coil Block Copolymer in the Presence of a Magnetic Field. *Macromolecules* **2011**, *44*, 7503–7507. [CrossRef]
97. Bras, W.; Emsley, J.W.; Levine, Y.K.; Luckhurst, G.R.; Seddon, J.M.; Timimi, B.A. Field-induced alignment of a smectic-A phase: A time-resolved X-ray diffraction investigation. *J. Chem. Phys.* **2004**, *121*, 4397–4413. [CrossRef] [PubMed]
98. Brimicombe, P.D.; Siemianowski, S.D.; Jaradat, S.; Levine, Y.K.; Thompson, P.; Bras, W.; Gleeson, H.F. Time-resolved x-ray studies of the dynamics of smectic-A layer realignment by magnetic fields. *Phys. Rev. E* **2009**, *79*, 031706. [CrossRef]
99. Torbet, J. Solution behavior of DNA studied with magnetically induced birefringence. *Methods Enzym.* **1992**, *211*, 518–532.
100. Torbet, J.; Dickens, M.J. Orientation of Skeletal-Muscle Actin in Strong Magnetic-Fields. *FEBS Lett.* **1984**, *173*, 403–406. [CrossRef]
101. Liebi, M.; van Rhee, P.G.; Christianen, P.C.M.; Kohlbrecher, J.; Fischer, P.; Walde, P.; Windhab, E.J. Alignment of Bicelles Studied with High-Field Magnetic Birefringence and Small-Angle Neutron Scattering Measurements. *Langmuir* **2013**, *29*, 3467–3473. [CrossRef] [PubMed]
102. Holmes, A.T.; Walsh, G.R.; Blackburn, E.; Forgan, E.M.; Savey-Bennett, M. A 17 T horizontal field cryomagnet with rapid sample change designed for beamline use. *Rev. Sci. Instrum.* **2012**, *83*, 023904. [CrossRef] [PubMed]
103. Linden, P.J.E.M.v.d.; Mathon, O.; Strohm, C.; Sikora, M. Miniature pulsed magnet system for synchrotron x-ray measurements. *Rev. Sci. Instrum.* **2008**, *79*, 075104. [CrossRef] [PubMed]
104. Duc, F.; Fabrèges, X.; Roth, T.; Detlefs, C.; Frings, P.; Nardone, M.; Billette, J.; Lesourd, M.; Zhang, L.; Zitouni, A.; et al. A 31 T split-pair pulsed magnet for single crystal x-ray diffraction at low temperature. *Rev. Sci. Instrum.* **2014**, *85*, 053905. [CrossRef] [PubMed]
105. Frings, P.; Vanacken, J.; Detlefs, C.; Duc, F.; Lorenzo, J.E.; Nardone, M.; Billette, J.; Zitouni, A.; Bras, W.; Rikken, G.L.J.A. Synchrotron x-ray powder diffraction studies in pulsed magnetic fields. *Rev. Sci. Instrum.* **2006**, *77*, 063903. [CrossRef]
106. Duc, F.; Tonon, X.; Billette, J.; Rollet, B.; Knafo, W.; Bourdarot, F.; Béard, J.; Mantegazza, F.; Longuet, B.; Lorenzo, J.E.; et al. 40-Tesla pulsed-field cryomagnet for single crystal neutron diffraction. *Rev. Sci. Instrum.* **2018**, *89*, 053905. [CrossRef] [PubMed]
107. Vesperinas, A.; Eastoe, J.; Wyatt, P.; Grillo, I.; Heenan, R.K.; Richards, J.M.; Bell, G.A. Photoinduced phase separation. *J. Am. Chem. Soc.* **2006**, *128*, 1468–1469. [CrossRef]
108. Tabor, R.F.; Oakley, R.J.; Eastoe, J.; Faul, C.F.J.; Grillo, I.; Heenan, R.K. Reversible light-induced critical separation. *Soft Matter* **2009**, *5*, 78–80. [CrossRef]
109. Oh, H.; Ketner, A.M.; Heymann, R.; Kesselman, E.; Danino, D.; Falvey, D.E.; Raghavan, S.R. A simple route to fluids with photo-switchable viscosities based on a reversible transition between vesicles and wormlike micelles. *Soft Matter* **2013**, *9*, 5025–5033. [CrossRef]
110. Morgan, B.; Dadmun, M.D. The role of incident light intensity, wavelength, and exposure time in the modification of conjugated polymer structure in solution. *Eur. Polym. J.* **2017**, *89*, 272–280. [CrossRef]
111. Morgan, B.; Dadmun, M.D. Illumination of Conjugated Polymer in Solution Alters Its Conformation and Thermodynamics. *Macromolecules* **2016**, *49*, 3490–3496. [CrossRef]
112. Bras, W.; Stanley, H. Unexpected effects in non crystalline materials exposed to X-ray radiation. *J. Non-Cryst. Solids* **2016**, *451*, 153–160. [CrossRef]
113. Liberton, M.; Page, L.E.; O'Dell, W.B.; O'Neill, H.; Mamontov, E.; Urban, V.S.; Pakrasi, H.B. Organization and Flexibility of Cyanobacterial Thylakoid Membranes Examined by Neutron Scattering. *J. Biol. Chem.* **2013**, *288*, 3632–3640. [CrossRef]
114. Perera, S.; Chawla, U.; Shrestha, U.R.; Bhowmik, D.; Struts, A.V.; Qian, S.; Chu, X.Q.; Brown, M.F.J. Small-Angle Neutron Scattering Reveals Energy Landscape for Rhodopsin Photoactivation. *Phys. Chem. Lett.* **2018**, *9*, 7064–7071. [CrossRef] [PubMed]
115. Whittaker, A.G.; Harrison, A.; Oakley, G.S.; Youngson, I.D.; Heenan, R.K.; King, S.M. Preliminary experiments on apparatus for in situ studies of microwave-driven reactions by small angle neutron scattering. *Rev. Sci. Instrum.* **2001**, *72*, 173–176. [CrossRef]
116. Toolan, D.T.W.; Adlington, K.; Isakova, A.; Kalamiotis, A.; Mokarian-Tabari, P.; Dimitrakis, G.; Dodds, C.; Arnold, T.; Terrill, N.J.; Bras, W.; et al. Selective molecular annealing: In situ small angle X-ray scattering study of microwave-assisted annealing of block copolymers. *Phys. Chem. Chem. Phys.* **2017**, *19*, 20412–20419. [CrossRef]
117. Greaves, G.N.; Wilding, M.C.; Fearn, S.; Langstaff, D.; Kargl, F.; Cox, S.; Van, Q.V.; Majerus, O.; Benmore, C.J.; Weber, R.; et al. Detection of first-order liquid/liquid phase transitions in yttrium oxide-aluminum oxide melts. *Science* **2008**, *322*, 566–570. [CrossRef]
118. Hennet, L.; Cristiglio, V.; Kozaily, J.; Pozdnyakova, I.; Fischer, H.E.; Bytchkov, A.; Drewitt, J.W.E.; Leydier, M.; Thiaudiere, D.; Gruner, S.; et al. Aerodynamic levitation and laser heating: Applications at synchrotron and neutron sources. *Eur. Phys. J. -Spec. Top.* **2011**, *196*, 151–165. [CrossRef]
119. Weber, J.K.R.; Rey, C.A.; Neuefeind, J.; Benmore, C.J. Acoustic levitator for structure measurements on low temperature liquid droplets. *Rev. Sci. Instrum.* **2009**, *80*, 083904. [CrossRef]
120. Cristiglio, V.; Grillo, I.; Fomina, M.; Wien, F.; Shalaev, E.; Novikov, A.; Brassamin, S.; Refregiers, M.; Perez, J.; Hennet, L. Combination of acoustic levitation with small angle scattering techniques and synchrotron radiation circular dichroism. Application to the study of protein solutions. *Biochim. Biophys. Acta-Gen. Subj.* **2017**, *1861*, 3693–3699. [CrossRef] [PubMed]

121. Krauss, S.W.; Schweins, R.; Magerl, A.; Zobel, M. Free-film small-angle neutron scattering: A novel container-free in situ sample environment with minimized H/D exchange. *J. Appl. Crystallogr.* **2019**, *52*, 284–288. [CrossRef] [PubMed]
122. Mitchell, J.B.A.; Le Garrec, J.L.; Florescu-Mitchell, A.I.; di Stasio, S. Small-angle neutron scattering study of soot particles in an ethylene-air diffusion flame. *Combust. Flame* **2006**, *145*, 80–87. [CrossRef]
123. Lee, Y.L.; Ristic, R.I.; DeMatos, L.L.; Martin, C.M. Crystallisation Pathways of Polymorphic Triacylglycerols Induced by Mechanical Energy. *XIV Int. Conf. Small-Angle Scatt.* **2010**, *247*, 012049. [CrossRef]
124. Gupta, S.; Bleuel, M.; Schneider, G.J. A new ultrasonic transducer sample cell for in situ small-angle scattering experiments. *Rev. Sci. Instrum.* **2018**, *89*, 7. [CrossRef] [PubMed]
125. Greenspan, L. Humidity fixed-points of binary saturated aqueous-solutions. *J. Res. Natl. Bur. Stand. Sect. A-Phys. Chem.* **1977**, *81*, 89–96. [CrossRef]
126. Kim, M.H.; Glinka, C.J.; Carter, R.N. In situ vapor sorption apparatus for small-angle neutron scattering and its application. *Rev. Sci. Instrum.* **2005**, *76*, 10. [CrossRef]
127. Plaza, N.Z.; Pingali, S.V.; Qian, S.; Heller, W.T.; Jakes, J.E. Informing the improvement of forest products durability using small angle neutron scattering. *Cellulose* **2016**, *23*, 1593–1607. [CrossRef]
128. Qian, S.; Rai, D.K. Grazing-Angle Neutron Diffraction Study of the Water Distribution in Membrane Hemifusion: From the Lamellar to Rhombohedral Phase. *J. Phys. Chem. Lett.* **2018**, *9*, 5778–5784. [CrossRef] [PubMed]
129. Yu, D.J.; An, K.; Gao, C.Y.; Heller, W.T.; Chen, X. A portable hydro-thermo-mechanical loading cell for in situ small angle neutron scattering studies of proton exchange membranes. *Rev. Sci. Instrum.* **2013**, *84*, 6. [CrossRef]
130. Putra, A.; Iwase, H.; Yamaguchi, D.; Koizumi, S.; Maekawa, Y.; Matsubayashi, M.; Hashimoto, T. In-situ observation of dynamic water behavior in polymer electrolyte fuel cell by combined method of Small-Angle Neutron Scattering and Neutron Radiography. *J. Phys. Conf. Ser.* **2010**, *247*, 012044. [CrossRef]
131. Koizumi, S.; Ueda, S.; Ananda, P.; Tsutsumi, Y. Heterogeneous cell performance of polymer electrolyte fuel cell at high current operation: Respiration mode as non-equilibrium phenomenon. *Aip Adv.* **2019**, *9*, 9. [CrossRef]
132. Ueda, S.; Koizumi, S.; Tsutsumi, Y. Initial conditioning of a polymer electrolyte fuel cells: The relationship between microstructure development and cell performance, investigated by small-angle neutron scattering. *Results Phys.* **2019**, *12*, 1871–1879. [CrossRef]
133. Jafta, C.J.; Sun, X.G.; Veith, G.M.; Jensen, G.V.; Mahurin, S.M.; Paranthaman, M.P.; Dai, S.; Bridges, C.A. Probing microstructure and electrolyte concentration dependent cell chemistry via operando small angle neutron scattering. *Energy Environ. Sci.* **2019**, *12*, 1866–1877. [CrossRef]
134. Hattendorff, J.; Seidlmayer, S.; Gasteiger, H.A.; Gilles, R. Li-ion half-cells studied operando during cycling by small-angle neutron scattering. *J. Appl. Crystallogr.* **2020**, *53*, 210–221. [CrossRef]
135. Djire, A.; Pande, P.; Deb, A.; Siegel, J.B.; Ajenifujah, O.T.; He, L.L.; Sleightholme, A.E.; Rasmussen, P.G.; Thompson, L.T. Unveiling the pseudocapacitive charge storage mechanisms of nanostructured vanadium nitrides using in-situ analyses. *Nano Energy* **2019**, *60*, 72–81. [CrossRef]
136. Bras, W.; Ryan, A.J. Sample environments and techniques combined with Small Angle X-ray Scattering. *Adv. Colloid Interface Sci.* **1998**, *75*, 1–43. [CrossRef]
137. Bras, W.; Koizumi, S.; Terrill, N.J. Beyond simple small-angle X-ray scattering: Developments in online complementary techniques and sample environments. *IUCrJ* **2014**, *1*, 478–491. [CrossRef]
138. Pullen, S.A.; Booth, N.; Olsen, S.R.; Day, B.; Franceschini, F.; Mannicke, D.; Gilbert, E.P. Design and implementation of a differential scanning calorimeter for the simultaneous measurement of small angle neutron scattering. *Meas. Sci. Technol.* **2014**, *25*, 8. [CrossRef]
139. Scalambra, F.; Rudic, S.; Romerosa, A. Molecular Insights into Bulk and Porous kappa P-2,N-PTA Metal-Organic Polymers by Simultaneous Raman Spectroscopy and Inelastic Neutron Scattering. *Eur. J. Inorg. Chem.* **2019**, *8*, 1155–1161. [CrossRef]
140. Adams, M.A.; Parker, S.F.; Fernandez-Alonso, F.; Cutler, D.J.; Hodges, C.; King, A. Simultaneous neutron scattering and Raman scattering. *Appl. Spectrosc.* **2009**, *63*, 727–732. [CrossRef] [PubMed]
141. Lopez-Barron, C.R.; Zeng, Y.M.; Schaefer, J.J.; Eberle, A.P.R.; Lodge, T.P.; Bates, F.S. Simultaneous Neutron Scattering and Raman Scattering. *Macromolecules* **2017**, *50*, 3627–3636.
142. Bryant, G.K.; Gleeson, H.F.; Ryan, A.J.; Fairclough, J.P.A.; Bogg, D.; Goossens, J.G.P.; Bras, W. Raman spectroscopy combined with small angle x-ray scattering and wide angle x-ray scattering as a tool for the study of phase transitions in polymers. *Rev. Sci. Instrum.* **1998**, *69*, 2114. [CrossRef]
143. Wurm, A.; Soliman, R.; Goossens, J.G.P.; Bras, W.; Schick, C. Evidence of pre-crystalline-order in super-cooled polymer melts revealed from simultaneous dielectric spectroscopy and SAXS. *J. Non-Cryst. Solids* **2005**, *351*, 2773–2779. [CrossRef]
144. Jimenez-Ruiz, M.; Sanz, A.; Nogales, A.; Ezquerra, T.A. Experimental setup for simultaneous measurements of neutron diffraction and dielectric spectroscopy during crystallization of liquids. *Rev. Sci. Instrum.* **2005**, *76*, 043901. [CrossRef]
145. Richards, J.J.; Wagner, N.J.; Butler, P.D. A strain-controlled RheoSANS instrument for the measurement of the microstructural, electrical, and mechanical properties of soft materials. *Rev. Sci. Instrum.* **2017**, *88*, 10. [CrossRef] [PubMed]
146. Peterson, P.F.; Campbell, S.I.; Reuter, M.A.; Taylor, R.J.; Zikovsky, J. Event-based processing of neutron scattering data. *Nucl. Instrum. Methods Phys. Res. Sect. A-Accel. Spectrometers Detect. Assoc. Equip.* **2015**, *803*, 24–28. [CrossRef]
147. Kaneko, F.; Seto, N.; Sato, S.; Radulescu, A.; Schiavone, M.M.; Allgaier, J.; Ute, K. Development of a Simultaneous SANS/FTIR Measuring System. *Chem. Lett.* **2015**, *44*, 497–499. [CrossRef]

148. Kaneko, F.; Seto, N.; Sato, S.; Radulescu, A.; Schiavone, M.M.; Allgaier, J.; Ute, K. Simultaneous small-angle neutron scattering and Fourier transform infrared spectroscopic measurements on cocrystals of syndiotactic polystyrene with polyethylene glycol dimethyl ethers. *J. Appl. Crystallogr.* **2016**, *49*, 1420–1427. [CrossRef] [PubMed]
149. Bras, W.; Derbyshire, G.E.; Bogg, D.; Cooke, J.; Elwell, M.J.; Komanschek, B.U.; Naylor, S.; Ryan, A.J. Simultaneous Studies of Reaction Kinetics and Structure Development in Polymer Processing. *Science* **1995**, *267*, 996–999. [CrossRef] [PubMed]
150. Hongladarom, K.; Ugaz, V.M.; Cinader, D.K.; Burghardt, W.R.; Quintana, J.P.; Hsiao, B.S.; Dadmun, M.D.; Hamilton, W.A.; Butler, P.D. X-ray scattering, and neutron scattering measurements of molecular orientation in sheared liquid crystal polymer solutions. *Macromolecules* **1996**, *29*, 5346–5355. [CrossRef]
151. Fernandez-Ballester, L.; Gough, T.; Meneau, F.; Bras, W.; Ania, F.; Balta-Calleja, F.J.; Kornfield, J.A. Simultaneous birefringence, small-and wide-angle X-ray scattering to detect precursors and characterize morphology development during flow-induced crystallization of polymers. *J. Synchrotron Radiat.* **2008**, *15*, 185–190. [CrossRef]
152. Lo Celso, F.; Triolo, A.; Triolo, F.; Thiyagarajan, P.; Amenitsch, H.; Steinhart, M.; Kriechbaum, M.; DeSimone, J.M.; Triolo, R. A combined small-angle neutron and X-ray scattering study of block copolymers micellisation in supercritical carbon dioxide. *J. Appl. Crystallogr.* **2003**, *36*, 660–663. [CrossRef]
153. Mao, Y.M.; Liu, K.; Zhan, C.B.; Geng, L.H.; Chu, B.; Hsiao, B.S. Characterization of Nanocellulose Using Small-Angle Neutron, X-ray, and Dynamic Light Scattering Techniques. *J. Phys. Chem. B* **2017**, *121*, 1340–1351. [CrossRef]
154. Liu, D.; Li, X.Y.; Song, H.T.; Wang, P.C.; Chen, J.; Tian, Q.; Sun, L.W.; Chen, L.; Chen, B.; Gong, J.; et al. Hierarchical structure of MWCNT reinforced semicrystalline HDPE composites: A contrast matching study by neutron and X-ray scattering. *Eur. Polym. J.* **2018**, *99*, 18–26. [CrossRef]
155. Metwalli, E.; Götz, K.; Lages, S.; Bär, K.; Zech, T.; Noll, D.M.; Schuldes, I.; Schindler, T.; Prihoda, A.; Lang, H.; et al. A novel experimental approach for nanostructure analysis: Simultaneous small-angle X-ray and neutron scattering. *arXiv* **2020**, arXiv:2003.12585v1. [CrossRef]
156. Romer, S.; Urban, C.; Lobaskin, V.; Scheffold, F.; Stradner, A.; Kohlbrecher, J.; Schurtenberger, P. Simultaneous light and small-angle neutron scattering on aggregating concentrated colloidal suspensions. *J. Appl. Crystallogr.* **2003**, *36*, 1–6. [CrossRef]
157. Doutch, J.; Bason, M.; Franceschini, F.; James, K.; Clowes, D.; Gilbert, E.P. Structural changes during starch pasting using simultaneous Rapid Visco Analysis and small-angle neutron scattering. *Carbohydr. Polym.* **2012**, *88*, 1061–1071. [CrossRef]
158. Lopez, C.G.; Watanabe, T.; Martel, A.; Porcar, L.; Cabral, J.T. Microfluidic-SANS: Flow processing of complex fluids. *Sci. Rep.* **2015**, *5*, 7727. [CrossRef]
159. Noda, I. Frontiers of Two-Dimensional Correlation Spectroscopy. Part 1. New concepts and noteworthy developments. *J. Mol. Struct.* **2014**, *1069*, 3–22.
160. Noda, I. Frontiers of two-dimensional correlation spectroscopy. Part 2. Perturbation methods, fields of applications, and types of analytical probes. *J. Mol. Struct.* **2014**, *1069*, 23–49. [CrossRef]
161. Smirnova, D.S.; Kornfield, J.A.; Lohse, D.J. Morphology Development in Model Polyethylene via Two-Dimensional Correlation Analysis. *Macromolecules* **2011**, *44*, 6836–6848. [CrossRef]
162. Haas, S.; Plivelic, T.S.; Dicko, C. Combined SAXS/UV-vis/Raman as a Diagnostic and Structure Resolving Tool in Materials and Life Sciences Applications. *J. Phys. Chem. B* **2014**, *118*, 2264–2273. [CrossRef] [PubMed]
163. Willendrup, P.K.; Udby, L.; Knudsen, E.; Farhi, E.; Lefmann, K. Using McStas for modelling complex optics, using simple building bricks. *Nucl. Instrum. Methods Phys. Res. Sect. A* **2011**, *634*, S150–S155. [CrossRef]
164. Lin, J.Y.Y.; Smith, H.L.; Granroth, G.E. MCViNE—An object oriented Monte Carlo neutron ray tracing simulation package. *Nucl. Instrum. Methods Phys. Res. Sect. A* **2016**, *810*, 86–96. [CrossRef]
165. Farhi, E.; Hugouvieux, V.; Johnson, M.; Kob, W. Virtual experiments: Combining realistic neutron scattering instrument and sample simulations. *J. Comput. Phys.* **2009**, *228*, 5251–5261. [CrossRef]
166. Hugouvieux, V.; Farhi, E.; Johnson, M.R. Structure and dynamics of le-G: Neutron scattering experiments and ab initio molecular dynamics simulations. *PRB* **2007**, *75*, 104208. [CrossRef]
167. Willendrup, P.; Filges, U.; Keller, L.; Farhi, E.; Lefmann, K. Validation of a realistic powder sample using data from DMC at PSI. *Phys. B Cond. Matt.* **2006**, *385*, 1032. [CrossRef]
168. Hexemer, A.; Müller-Buschbaum, P. Advanced grazing-incidence techniques for modern soft-matter materials analysis. *IUCrJ* **2015**, *2*, 106–125. [CrossRef] [PubMed]
169. Muller-Buschbaum, P. Grazing incidence small-angle neutron scattering: Challenges and possibilities. *Polym. J.* **2013**, *45*, 34–42. [CrossRef]

Article

A Unified User-Friendly Instrument Control and Data Acquisition System for the ORNL SANS Instrument Suite

Xingxing Yao, Blake Avery, Miljko Bobrek, Lisa Debeer-Schmitt, Xiaosong Geng, Ray Gregory, Greg Guyotte, Mike Harrington, Steven Hartman, Lilin He, Luke Heroux, Kay Kasemir, Rob Knudson, James Kohl, Carl Lionberger, Kenneth Littrell, Matthew Pearson, Sai Venkatesh Pingali, Cody Pratt, Shuo Qian *, Mariano Ruiz-Rodriguez, Vladislav Sedov, Gary Taufer, Volker Urban and Klemen Vodopivec

Oak Ridge National Laboratory, Oak Ridge, TN 37830, USA; xingxingyao@gmail.com (X.Y.);
averybe@ornl.gov (B.A.); bobrekm@ornl.gov (M.B.); debeerschmlm@ornl.gov (L.D.-S.); geng@ornl.gov (X.G.);
gregoryrd@ornl.gov (R.G.); guyottegs@ornl.gov (G.G.); harringtonml@ornl.gov (M.H.);
hartmansm@ornl.gov (S.H.); hel3@ornl.gov (L.H.); herouxla@ornl.gov (L.H.); kasemirk@ornl.gov (K.K.);
knudsoniroiv@ornl.gov (R.K.); kohlja@ornl.gov (J.K.); calionberger@lbl.gov (C.L.); littrellkc@ornl.gov (K.L.);
pearsonmr@ornl.gov (M.P.); pingalis@ornl.gov (S.V.P.); prattcl@ornl.gov (C.P.); ruizmm@ornl.gov (M.R.-R.);
sedovvn@ornl.gov (V.S.); tauferga@ornl.gov (G.T.); urbanvs@ornl.gov (V.U.); vodopiveck@ornl.gov (K.V.)
* Correspondence: qians@ornl.gov; Tel.: +1-865-241-1934

Abstract: In an effort to upgrade and provide a unified and improved instrument control and data acquisition system for the Oak Ridge National Laboratory (ORNL) small-angle neutron scattering (SANS) instrument suite—biological small-angle neutron scattering instrument (Bio-SANS), the extended q-range small-angle neutron scattering diffractometer (EQ-SANS), the general-purpose small-angle neutron scattering diffractometer (GP-SANS)—beamline scientists and developers teamed up and worked closely together to design and develop a new system. We began with an in-depth analysis of user needs and requirements, covering all perspectives of control and data acquisition based on previous usage data and user feedback. Our design and implementation were guided by the principles from the latest user experience and design research and based on effective practices from our previous projects. In this article, we share details of our design process as well as prominent features of the new instrument control and data acquisition system. The new system provides a sophisticated Q-Range Planner to help scientists and users plan and execute instrument configurations easily and efficiently. The system also provides different user operation interfaces, such as wizard-type tool Panel Scan, a Scripting Tool based on Python Language, and Table Scan, all of which are tailored to different user needs. The new system further captures all the metadata to enable post-experiment data reduction and possibly automatic reduction and provides users with enhanced live displays and additional feedback at the run time. We hope our results will serve as a good example for developing a user-friendly instrument control and data acquisition system at large user facilities.

Keywords: SANS; neutron scattering; instrument control; data acquisition; user facility; GUI

1. Introduction

Small-angle neutron scattering (SANS) is a powerful technique to resolve structures from a few to hundreds of nanometers in a wide range of materials. The SANS instrument suite at the Oak Ridge National Laboratory (ORNL) neutron scattering facilities—including the biological small-angle neutron scattering instrument (Bio-SANS), the general-purpose small-angle neutron scattering diffractometer (GP-SANS), and the extended q-range small-angle neutron scattering diffractometer (EQ-SANS)—serve many different research communities, including biology, soft matter, quantum materials, and metallurgy [1].

The three instruments are custom-developed and built with similar yet different components at two different neutron sources, the spallation neutron source (SNS) and the high flux isotope rector (HFIR) at ORNL. While there is much overlap among the instrument specifications, each of these instruments has unique advantages for different types of experiments. Thus, there is considerable overlap in their user bases, who use multiple instruments for different experimental needs, such as specific performance or sample environment equipment needs. The SANS instrument suite has had very different instrument control and data acquisition systems (IC-DASs) over the past decade for historical reasons. EQ-SANS, located at the SNS, relied on PyDAS, a Python extension of the original home-developed SNS DAS [2]. Bio-SANS and GP-SANS, both located at HFIR, were served by a customized graphic user interface (GUI) extension based on the Spectrometer Instrument Control Environment (SPICE) software built on LabView [3].

At the SNS, the original SNS DAS played a critical role in the commissioning and operation of early instruments. It also laid the foundation of the data format for timestamped event mode data coming from the pulsed neutron source, which is required to record the time-of-flight of each neutron for further data process and reduction [4,5]. Later instruments with higher data throughput were commissioned at the SNS and exposed the data acquisition bandwidth limitations of the original SNS DAS. Years of operational experience also highlighted the need to improve the reliability and usability of the original SNS DAS. As a result, a series of significant software and hardware upgrade projects have taken place at the SNS [6–9], using the Experimental Physics and Industrial Control System (EPICS) toolkit and Control System Studio (CS-Studio) [10–12] that are used by the SNS accelerator control system. EQ-SANS at the SNS was the first among the SANS suite to be upgraded to the EPICS-based IC-DAS.

Over time, significant upgrades at Bio-SANS and GP-SANS, including upgrades allowing them to use the same type of ^{3}He linear position-sensitive detectors used at EQ-SANS [13], neutron collimation systems, and additional sample environment equipment [1], have challenged the existing capabilities of SPICE and DAS. For example, the timestamped event mode data from the detector system have been converted to histograms to be handled by SPICE, losing valuable temporal information for neutron detection at Bio-SANS and GP-SANS. Moreover, given the growing suite of sample environment equipment shared among the SANS instruments and the need for a unified user experience, it is imperative that Bio-SANS and GP-SANS follow suit to upgrade to the EPICS-based system. This unification of the IC-DAS also will make system management and operation easier without splitting resources on what otherwise might be two separate development efforts.

This article shares our user needs and requirements analysis results, key methods we relied on, and new features of the EPICS-based SANS IC-DAS. Besides basic instrument control and data acquisition features, our new system is ready to be deployed for other similar SANS instruments, offering accurate and intuitive metadata handling, sophisticated experiment planning tools, and different user operation interfaces catering to different user communities.

2. User Needs and Requirements Analysis
2.1. Diverse User Needs

The users of the SANS IC-DAS are quite diverse. Categorized by different roles, they can be external facility users, instrument scientists, and supporting teams including sample environment teams, detector teams, and so forth. In addition to the rudimentary instrument control functionality, each role has more specific requirements of the system. For example, supporting teams require in-depth control of instrument components, clear status indication, and sufficient logs for monitoring and diagnostic troubleshooting. Instrument scientists have a good overall knowledge of the whole instrument system and are highly experienced in configuring an instrument for specific scientific requirements. However, they need to be freed from tedious hands-on operations to focus on scientific aspects of user experiments. The largest and most important group are external neutron scattering facility users. We want to empower them to successfully conduct experiments and collect useful measurement data with minimum effort expended on instrument operation. As the SANS instrument suite covers a large span of scientific communities with different skill sets; different experience levels with neutron scattering instruments; and, most crucially, different ways of conducting experiments suited for their sciences, our IC-DAS must make it easy for all of those communities to be productive during the short period of time they spend at an instrument. For example, users from the experimental condensed matter physics community usually have a single sample, but require various sample environment conditions in a particular experiment. They need to be able to easily visualize and interpret the reduced data to decide on the next condition to explore. They usually spend more time on a sample and require quick, flexible manipulation of many instrument and sample environment parameters. On the other hand, biochemists or biologists conducting solution SANS experiments usually use highly standardized instrument configurations for a series of samples with small variants. They need an effortless way of setting up their experiments, while they focus primarily on preparing fresh samples on-site. For these two extreme cases and others between them, a variety of user interfaces reflecting different workflows need to be designed and developed.

Our analysis of previous usage data further supports the above requirements. For example, the wizard-like GUI extension (Figure 1) previously developed within SPICE served 99.4% of Bio-SANS experiments and 7.8% of GP-SANS experiments over the 2 years since its deployment in 2016 (HFIR was in a long outage from late 2018 to late 2019). The rest of the experiments during that period were served by SPICE macros similar to scripting interfaces. Based on these findings, the new system implements and improves on both of these tools, offering wizard-like Panel Scans and Scripting Tool based on Python, in addition to the simple start/stop button-click GUI and Table Scan feature that were part of the previous EPICS/CS-Studio system.

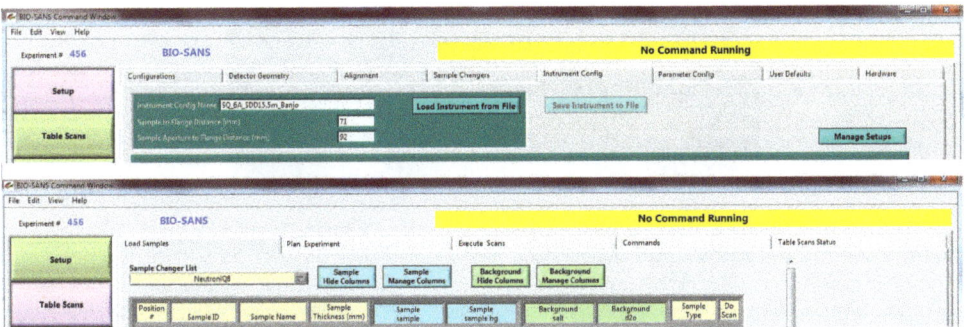

Figure 1. The Spectrometer Instrument Control Environment (SPICE) graphic user interface (GUI) extension with a wizard-like tool to set up scans for small-angle neutron scattering (SANS); highlighted in purple is the currently selected function group.

2.2. Complex Data and Metadata

During a neutron scattering experiment, an extensive amount of data and metadata are collected. The data are mainly events detected by the large two-dimensional (2D) position-sensitive neutron detectors with timestamps of each neutron event referring to a reference time. Such event mode data provide convenience for post-collection time-slicing or synchronizing with sample environment parameters for time-resolved studies or a stroboscopic method. The metadata, also recorded with timestamps, include inherent instrument parameters such as hardware component positions, health conditions, and sample environment equipment readouts, and user-input supplementary information such as sample and background details. All of those parameters are critical for setting up the correct instrument measurement configuration, and many of them are required for correct data reduction and interpretation.

Most inherent parameters are motor positions and other equipment readouts (Figure 2). They can be grouped in relation to (1) beam characteristics (e.g., status of neutron guides, source aperture, attenuator, beam trap); (2) the Q-range (the momentum transfer, Q, is the most important factor in a small-angle scattering experiment, representing the size range to be probed in the reciprocal space within which SANS measures; e.g., velocity selector rotation speed and tilt angle, sample aperture size, detector motor positions); (3) sample environment devices and conditions (e.g., temperature, magnetic field, pressure, rotating cell speed); and (4) sample changer and slot number. Individual motor position and other readings are difficult for both expert and non-expert users to comprehend; therefore, it is necessary to associate and display them with more meaningful configuration descriptions. In addition, a higher-level collective setup can be used to configure an instrument without setting those parameters individually.

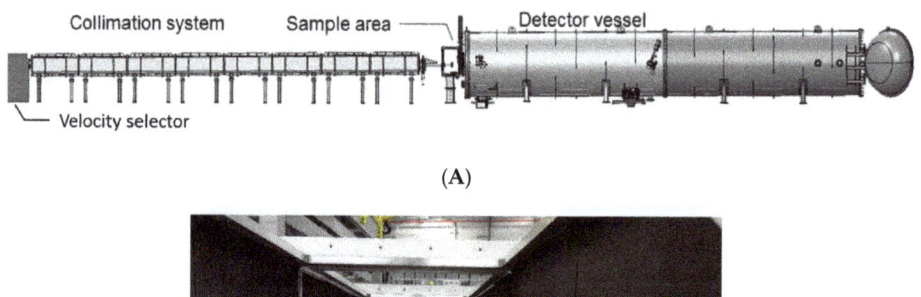

Figure 2. (**A**) An overview of a SANS instrument system. (**B**) The collimation system used at the biological small-angle neutron scattering instrument (Bio-SANS) and the general-purpose small-angle neutron scattering diffractometer (GP-SANS) with interchangeable neutron guide and aperture systems. There are eight sections of almost identical units with independent motor control.

User-input supplementary information is needed for users to keep track of different parameters such as slight variations in composition, concentration, matching background, and so on. Some of this information can be pulled from the centralized sample management database, inventory tracking of equipment, material and sample (ITEMS), whereas some can be provided only by users manually during an experiment. Moreover, the system needs to provide an expandable pathway for adopting emerging standards for metadata, such as Information System for Protein crystallography Beamlines (ISPyB) used by the biological small-angle scattering community [14] and the collective action for nomadic small-angle scatterers (canSAS, www.cansas.org) community.

2.3. Integrated Experiment Planning Tool

Our instrument scientists and users have been using simple calculating tools such as Excel spreadsheets to plan experiments, including Q-range and measurement time, and manually convert that information into actual instrument parameters to use. Given the increasing complexity of instruments—e.g., multiple detectors, different sample changers, multiple beam stops, sample environment devices—an integrated experiment planning tool that can take advantage of the known constraints of the parameters, and then easily save the planned configuration for use and reuse, is valuable. Clearly defined instrument configurations will also enable the implementation of automatic data reduction and provide coarse real-time data reduction, which is very helpful to guide users during measurements.

3. Methods

3.1. Needs-Driven and User-Centered Design

The developers initialized a thorough user experience–focused study on the existing SPICE and other relevant software environments. The study was guided by the principles from the latest user experience and design thinking research (e.g., references [15–17]), and based on effective practices from previous EPICS upgrade projects, such as using a beginner mindset, maintaining operational flexibility, and balancing between overall performance and individual process optimization. With help from database administrators, the previous metadata from SPICE (already ingested into the catalog database) were used to mine useful usage data to quantify the findings of our study. This effort not only helped clarify the required functionality but also helped prioritize the requirements based on evidence rather than impressions. The study further helped frame a shared vision of delivering an IC-DAS that is both functional and easy-to-use and helped build a relationship of trust between instrument scientists and control system developers.

Based on a goal of minimizing the physical, mental, and emotional efforts required of users in carrying out their tasks, the study identified four focus areas that could potentially be improved for a better user experience. These four focus areas are (1) the Q-range configuration representing the instrument configuration, including many component settings; (2) a user interface including both wizard-like Panel Scans and a Python-based Scripting Tool; (3) customizable detailed sample and buffer information tracking; and (4) detector geometry handling with various viable offsets and motor positions. With these in focus, we co-designed all main components of our high-level user-oriented tools; the details are provided in the results section.

3.2. Automation Based on Process Knowledge

Following the principle of "first make it work, then make it better," we built on the team's expertise and experience to create shared process knowledge regarding how the different parts of the SANS instruments are interwoven with one another. For example, as mentioned earlier, setting an instrument to a specific Q-range configuration requires coordinating the establishment of the beam characteristics with the settings for other hardware motors. For instance, the collimator guides and apertures used to define the incident beam are coupled with several pieces of distance metadata for Q-range calculation and other settings. This interwoven nature of the instrument parameters should be considered in aspects including instrument control, experiment planning, and metadata handling.

Another challenge is effectively managing instrument Q-range configurations with small variations that are sample- and proposal-specific. Improved understanding of the process knowledge in actual operation enabled us to differentiate between configurations that are standard or are changed infrequently, and those that are changed often. We did so by creating a sample- and proposal-dependent layer that allows a user to save Q-range configurations with only minor modifications, while still sharing the parameter files (e.g., flood, beam center, and dark current files) associated with the standard configurations. We designed our tools with as much automation as possible and set the order of development based on shared process knowledge.

4. Results

Within the user needs and requirements scope identified earlier, built on the EPICS system architecture (Figure 3), we developed and delivered a distributed IC-DAS customized for the SNS and HFIR SANS instruments. We added customizations and improvements to already developed features and building blocks, significantly reducing the user experience gaps encountered among the three SANS instruments at ORNL. This system is built on the abundant base of EPICS drivers that interface with different items of physical hardware, such as motors, temperature controllers, and magnets. Among modular applications (also referred to as input–output controllers) and different user interfaces, a scan server application serves as the "brain" and controls the overall instrument state while maintaining a "command queue" for future measurements. Various user interfaces (including Panel Scans and Scripting Tool) cater to different needs and preferences and enable users to plan and conduct their experiments efficiently. This architecture ensures a sound and flexible system that can meet the requirements of complex instruments such as the SANS instrument suite.

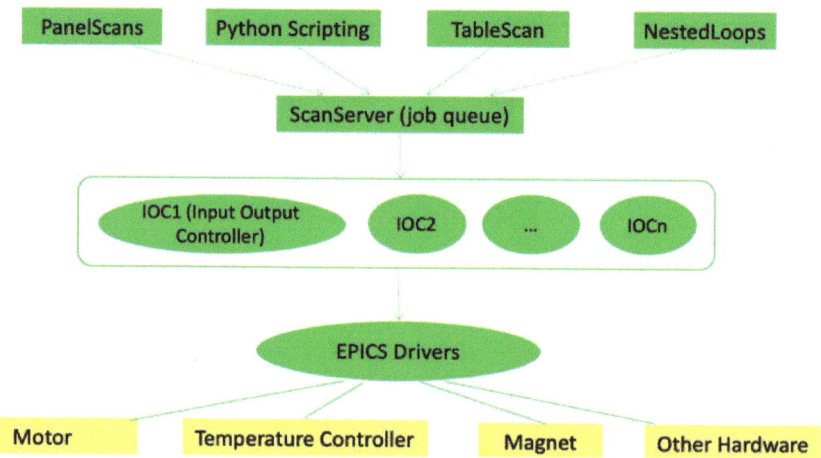

Figure 3. Experimental Physics and Industrial Control System (EPICS) architecture (NestedLoops is a generic graphic user interface tool that was previously developed to build scan jobs with simple nested loops logic. It aims for general use and thus retains a simple design. It can be commissioned if needed.).

In this section, we detail a few novel aspects of our system and discuss how they provide users with an improved instrument control and data handling experience. Additional screenshots of the new system are exhibited in the Supplementary Materials.

4.1. Intuitive Motion Control with Reliable Metadata

Following a good practice implemented in the previous SPICE software, we grouped the control of multiple devices into pseudo-motors. For example, for collimator motion control with eight different sections of guide motion control, a single "nguides (NGuides)" (number of guides) command can coordinate the control of guide motions in and out to keep a specific number of guides in the beam, along with apertures in the beam configuration. Note that the number of guides usually define the apparent source-to-sample distance (the flight path from the end of last guide to sample position). The distance needs to be coordinated with the sample-to-detector distance, various aperture sizes, and detector pixel resolution to provide optimal scattering geometry in SANS [18]. We included additional enhancements such as the newly defined 20 mm aperture in the beam configuration (aka, NGuides = "−1" in Figure 4); automated motor homing procedures; and logic embedded within the collimator motion control software to consistently compute and update the source aperture diameter and the metadata for the source-to-sample distance, based on motor positions. Similarly, detailed detector distance calculation logic is also integrated within the motion control to simplify or automate various offsets (e.g., sample-to–silicon window offset and detector motor position, see Figure 5) with a combined pseudo-motor total sample-to-detector distance. The latter matches the typical convention employed by SANS users in detector geometry and is critical in experiment planning and data reduction. The pseudo-motor is controllable like any physical motor, and the corresponding actual motor position is calculated based on offset values that can be changed according to different experiment setups. Note that all individual values and combined pseudo-motor values are captured in metadata redundantly in case they need to be cross-checked. The development and testing effort in motion control has been rewarded by a clean, simplified interface with enhanced functionalities, tighter integration with high-level tools, and more reliable metadata.

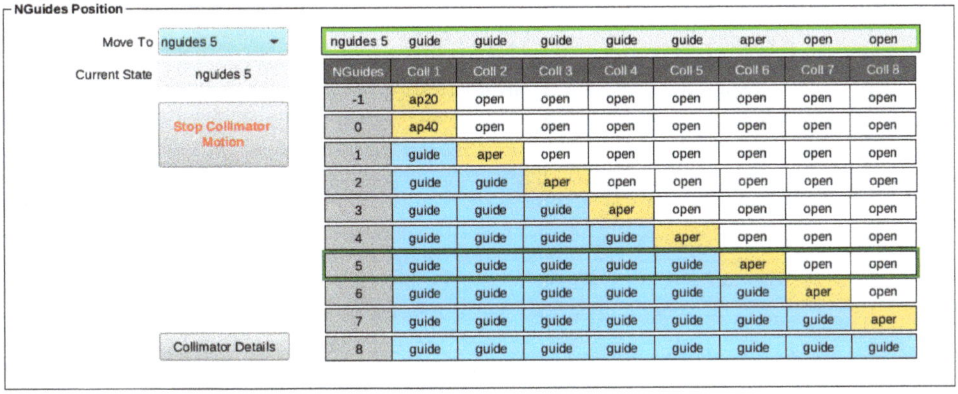

NGuides	Coll 1	Coll 2	Coll 3	Coll 4	Coll 5	Coll 6	Coll 7	Coll 8
nguides 5	guide	guide	guide	guide	guide	aper	open	open
-1	ap20	open	open	open	open	open	open	open
0	ap40	open	open	open	open	open	open	open
1	guide	aper	open	open	open	open	open	open
2	guide	guide	aper	open	open	open	open	open
3	guide	guide	guide	aper	open	open	open	open
4	guide	guide	guide	guide	aper	open	open	open
5	guide	guide	guide	guide	guide	aper	open	open
6	guide	guide	guide	guide	guide	guide	aper	open
7	guide	guide	guide	guide	guide	guide	guide	aper
8	guide	guide	guide	guide	guide	guide	guide	guide

Figure 4. An example of motor grouping, the pseudo-motors for guide operation in the collimation system.

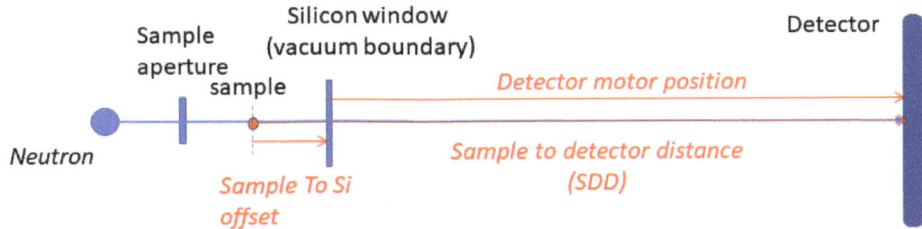

Figure 5. The diagram shows that the sample-to-detector distance combines the sample-to–silicon window offset and the actual detector motor position to form a controllable pseudo-motor.

4.2. A Customized and Fully Integrated Q-Range Planner

Instrument-specific experiment planning is important for the success of an experiment. For SANS experiments, Q-range is one of the most critical factors, as it determines the size range of a measurement. Previously, instrument scientists developed spreadsheets or other calculation tools independently to do their planning without much instrument-specific information or constraints. The integration of the Q-Range Planning tool within the IC-DAS enables the direct transfer of the Q-range configuration from the planning stage to the measurement stage. In addition to the benefit of imposing the physical constraints (e.g., motor limits) of a specific instrument, this implementation reduces the number of Q-range configurations that are due to small, unnecessary inconsistencies.

Once we understood these requirements, a customized Q-Range Planner was developed based on instrument scientists' spreadsheet calculators, as well as previous work on other instruments. The SANS Q-Range Planner helps users specify/update factors such as wavelength, attenuation factor, number of guides, aperture sizes, detector distance/rotation, distance offsets, and beam trap configuration for both scattering and transmission measurements. The factors then are converted into actual hardware settings such as motor positions. The planner then calculates the minimum Q at the beam stop rim, depending on the beam stop chosen, maximum Q values at corners and edges on each detector, and direct beam size on the detector; and, when applicable, it calculates the overlapping ratios between different detectors. Users can easily save a Q-range configuration to use and reuse in actual measurements. The saved Q configuration is in a human-readable text file format.

The SANS Q-Range Planner is deployed among the instrument suite. At different instruments, the calculation incorporates different instrument component details (e.g., detector, collimator, and beam trap details) but with an almost identical high-level interface for users (Figure 6). We also added beam center enhancement so users could more accurately calculate Q-ranges based on the specified beam center on the 2D display, instead of always pretending the beam center is at the previous physical detector center. The beam center coordination and calculated Q-range details are captured in each Q configuration file, reused at the run time for live displays (see Section 4.4), and saved in each data file to be used by the data reduction software. To meet the different needs and access privileges of external users and instrument scientists, two instances of the Q-Range Planner are running on each instrument. One is dedicated to use by instrument scientists to establish new standard configurations. The other is embedded within Panel Scans (see Section 4.3) to flexibly deal with sample- or experiment-dependent minor modifications, as well as to collect data with more than one Q-range configuration. Q-range configurations can also be easily set like a simple variable by using other user operation interfaces such as Scripting Tool and the generic Table Scan. With Q-Range configurations generated by the planner, a soft matter simulator (Figure S7) has been developed for generating typical model scattering curves with realistic instrument factors such as flux at different neutron guide numbers, Q resolutions, etc., to further aid users to plan an experiment.

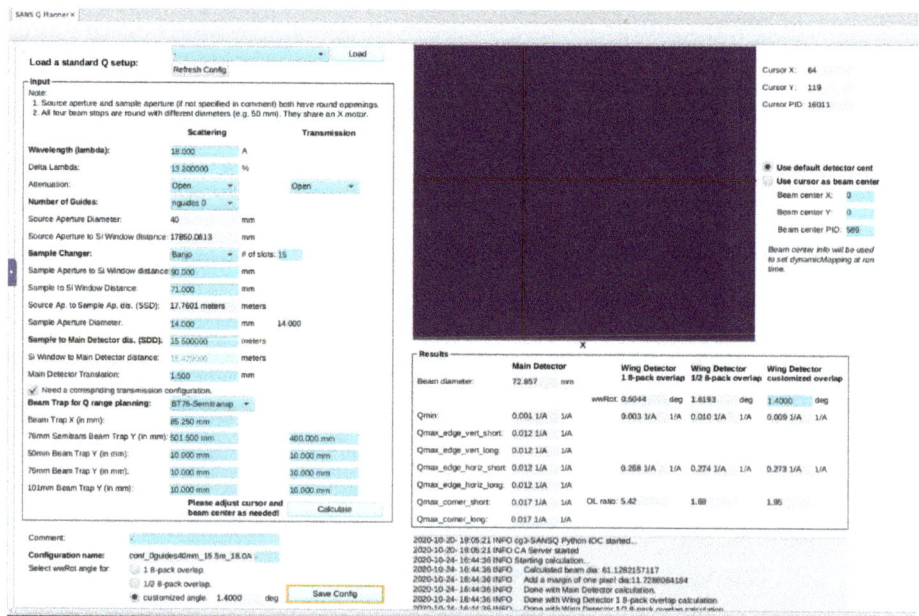

Figure 6. The Q-Range Planner.

4.3. User Operation Interfaces: Panel Scans and Scripting Tool

The previous EPICS/CS-Studio system included a generic Table Scan feature that used a spreadsheet to set up a list of scans for batching measurement. However, the intuitiveness and ease of use of the generic Table Scan was limited. To provide more functional and straightforward interfaces to different users, we implemented a wizard-like GUI-based tool, Panel Scans, and a Scripting Tool with an improved scripting middle-layer library. Both can use the Q-range configurations generated from the Q-Range Planner introduced in the previous section.

The SANS Panel Scans tool was designed based on earlier development of the SPICE GUI extension and other EPICS high-level scanning tools. It uses one instance of the Q-Range Planner and a six-step workflow (Figure 7) to guide users through the complex measurement planning process. This includes (1) checking or modifying individual Q-range configurations; (2) selecting up to four Q configurations in any order; (3) configuring sample environment variables; (4) keeping track of sample- and buffer-related variables for each sample changer slot; (5) specifying the exposure time for each combination of sample changer slot/Q configuration/sample environment variables; and (6) expanding the entire batch with all the details into a tabulated scan list. The scan list can then be simulated and tested before execution, submitted to the scan server to execute, and saved as a file for future reference.

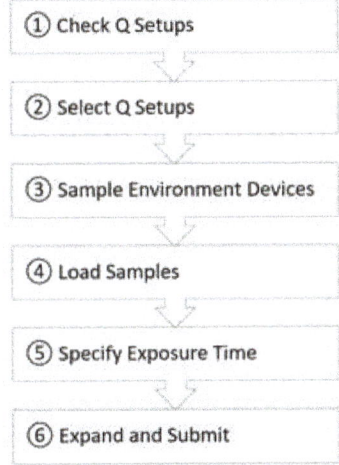

Figure 7. The workflow laid out by different tabs in the Panel Scans interface (details on the workflow tabs are presented in the Supplemental Materials).

The guided workflow and clear steps of Panel Scans not only make the complex process more manageable and less overwhelming for even new users, but also encapsulate many thoughtful features and improvements within an appropriate sub-task context. These features include a user-friendly table view and visualization of Q-ranges (see Figure 8), generation of scripting snippets with Q-range configurations selected (Figure 8), support for both pre-configured and ad-hoc sample environment variables for routine and innovative experiments, standardized sample- and buffer-related metadata variables (still with customizable tags for automatic background association), and accommodation of flexible expanding orders. The system also provides a high degree of freedom for users to set up and change their measurements. For example, if users choose to submit each table row as a separate scan job, they can use the upgraded scan monitor to reorder or delete and resubmit queued scan jobs. In addition, the "load sample" step provides customizable sample information to obligate users to keep track of variants among different samples, which encourages good practices for accurate supplementary information and enables future automatic background matching in data reduction.

Similarly, in Scripting Tool, the Scan Tools helper library also incorporates lessons learned from the legacy PyDAS, SPICE macros, and several early generations of home-developed EPICS Python middle layer libraries. It supports all Python features and additional scan-specific commands to be executed by scan server. The scientists worked together with the developers to ensure that the scripting library is easy to use for users with different levels of programming experience (see Figure 9). Other improvements that come with this implementation include converting other existing high-level scanning tools to use the same helper library, simplified and enhanced device configuration (including limits checking), and a simple screen for submitting scripts.

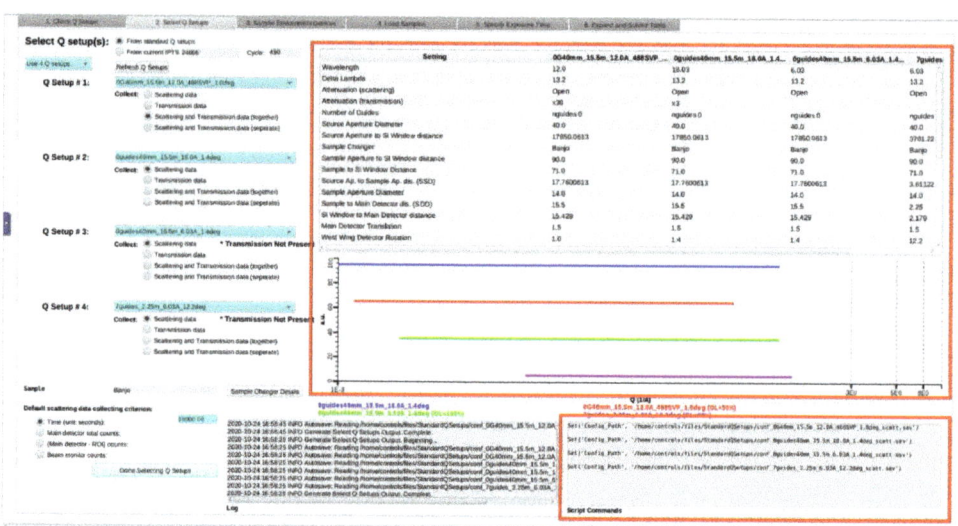

Figure 8. Friendly table view and visualization of Q-ranges (highlighted by the upper red rectangle box), scripting snippets generating with Q-range configurations selected (lower red box). The example shows up to four Q setups at the same time.

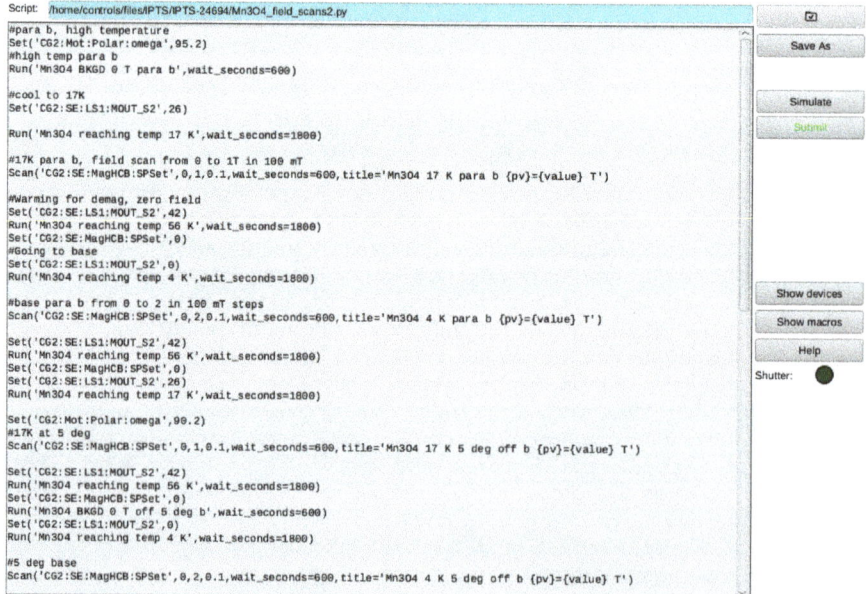

Figure 9. An example of the Python script used in Scripting Tool to control the magnet, cryostat temperature, and polarization for a highly flexible experiment.

4.4. Event Data Mode and New Features Related to Live Displays

One of the results of this upgrade is that it enables the production of event mode data for the whole DAS, especially the detectors at HFIR. Event mode imposes timestamps on neutron events detected and on other instrument systems, such as fast sample environment devices, enabling convenient post-data collection filtering. This is very useful in experiments that involve time-resolved kinetic studies, stroboscopic methods, setting up a time-of-flight source at HFIR with choppers, or simply detecting sample deterioration over time. At SNS, "event mode" refers to the pulsed accelerator timing signal; however, as a continuous neutron source, HFIR did not have an inherent timing signal as a reference. Therefore, a time server was set up for HFIR instruments; and Network Time Protocol [19] is used to synchronously timestamp events at an instrument, including neutron events and fast sample environment data, at a precision of within 1 ms. The system was made compatible with SNS event-based acquisition, which uses a timestamping rollover of 16.6667 ms, corresponding to the 60 Hz pulse rate. In other words, the events are grouped by their occurrence in the 16.6667 ms windows, and as such are saved in data files in Nexus format. For reactor-based instruments such as Bio-SANS and GP-SANS, the event timestamps are merely additional information about the neutron events and can be ignored if only the histogram of the data is used. No other changes or additions at HFIR SANS instruments are needed for event-based acquisition, as the relevant features had already been developed and tested on the SNS instruments. In Figure 10, an example of event mode data shows the rapid precipitation that occurs in a novel high strength, high ductility alloy that occurs as the sample was heated from room temperature to 700 °C from an experiment on GP-SANS [20]. The size and number density of the precipitates mediates the balance between strength and ductility in this family of alloys. The SANS data can detect such precipitates in-situ. This data was collected in the event mode in a single data set and was processed post-experimentally to split it up by 60-s interval during the temperature ramp and ten 300-s intervals for the annealing temperature. Moreover, the post-experiment process can be flexible depending on samples and sample conditions. The new capability also allows the visualization of sequential defined-time snapshots of the event mode data to aid real-time experimental evaluation and planning.

We have been using EPICS area detector driver for neutron event data (ADnED) and dynamicMapping (a Python tool for live data conversions) [9] to provide meaningful 2D and 1D live displays on other instruments. There are three related new features on SANS that may be worth mentioning. First, an ImageSnapshot Python tool has been deployed on the SANS instruments (and a few other instruments), which saves snapshots of histogrammed 2D detector images at the end of a run. These images are further parsed and ingested in the data catalog database for a quick view of measurement data. They can also be used for quick data reduction without reading the large HDF5 files if event mode data is not needed. Second, by updating dynamicMapping and saving and reusing beam center information in Q-range configurations, we were able to fully automate dynamicMapping to update mapping files for live d-spacing and Q conversions and for cursor display on the histogrammed 2D detector images. Third, a prototype Python tool, SANS-AutoRebin, has been developed to bin the data integrated by ADnED into log bins using default Q-range configurations, normalize log-binned data according to the number of pixels contributing to each log bin, and then scale the normalized data to beam monitor counts or run time. This prototype tool is part of our effort to provide users with live, coarsely reduced data for better steering of their experiments during the run time.

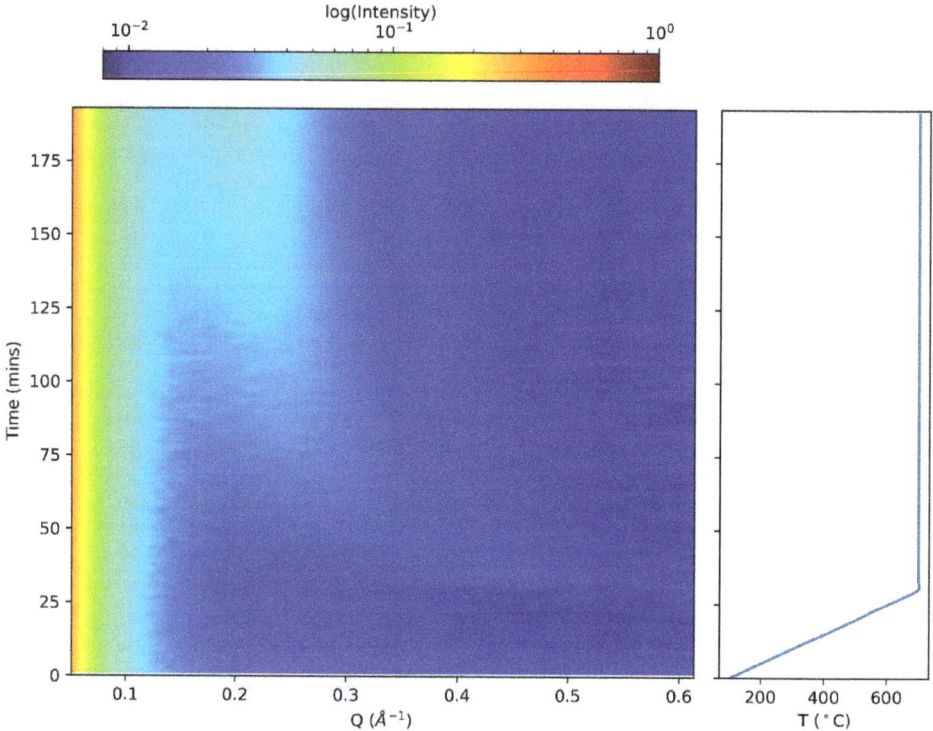

Figure 10. The SANS data processed over the time course of an experiment showing the rapid precipitation in a novel alloy under heating. On the left side, each slice along Q is a traditional Intensity vs. Q curve in SANS (the intensity is in cm^{-1}). On the right side is the temperature of the sample environment over time, also recorded as the event mode data.

5. Conclusions

Our new IC-DAS is based on EPICS, an open source distributed control system environment that has been widely adopted by other large scientific facilities. The systems completed commissioning at all SANS instruments by early 2020; an example data using the event mode is shown in Figure 10. With carefully designed and developed features to meet the needs of the SANS user communities, our new system provides a uniform user experience across the ORNL SANS neutron scattering instrument suite. Collaboratively, we designed and delivered a system that reflects a deeper understanding of our diverse user needs, instrument configurations, and complex experiment processes. Easy-to-use tools and more automation result in less cognitive load, confusion, stress, and human error for novice users, and more efficient use of beam time with higher-quality measurement data. The flexible architecture of the new system can handle the ever-increasing complexity of modern instruments. It also provides a manageable solution for integrating existing and new sample environment equipment across all SANS instruments. We hope the enhanced capabilities of our new system and the improved user experience it enables can also benefit other, similar instruments by improving operation for more productive scientific discovery.

Supplementary Materials: The following are available online at https://www.mdpi.com/2076-3417/11/3/1216/s1, Figure S1: Check Q Setups tab in the Panel Scans interface. The yellow outline highlights buttons are required to be clicked to ensure the output parameters to be calculated. Figure S2: Select Q Setups tab in the Panel Scans interface. Figure S3: Sample Environment Devices tab in the Panel Scans interface, for selecting specific sample environment for the current experiment. "Use other device combination:" will reveal the text input box to type a comma separated parameter names or aliases. Figure S4: Load Samples tab in the Panel Scans interface, for more specific sample information. Figure S5: Specify Exposure Time tab in the Panel Scans interface to setup measurement time or detector count at different configurations and samples. Figure S6: Expand and Submit tab in the Panel Scans interface. It expands the scans in different ways with all conditions from previous setups (such as samples, sample environment, configurations, measurement type (transmission, scattering or both)), only part of the columns are shown in the screenshot. Figure S7: The soft matter simulator with instrument specific parameters

Author Contributions: Conceptualization, X.Y., L.D.-S., R.G., G.G., S.H., L.H. (Lilin He), R.K., K.L., S.V.P., S.Q., and V.U.; funding acquisition, G.T.; investigation, M.B., L.D.-S., X.G., R.G., G.G., M.H., S.H., L.H. (Lilin He), L.H. (Luke Heroux), K.K., J.K., C.L., K.L., M.P., S.V.P., C.P., S.Q., M.R.-R., V.S., and K.V.; methodology, B.A., M.H., R.K., J.K., K.L. and, M.R.-R.; project administration, X.Y. and R.K.; resources, L.H. (Luke Heroux) and R.K.; software, X.Y., B.A., M.B., X.G., R.G., G.G., K.K., M.P., M.R.-R., V.S., G.T., and K.V.; supervision, S.H.; writing—original draft, X.Y. and S.Q.; writing—review and editing, X.Y. and S.Q. All authors have read and agreed to the published version of the manuscript.

Funding: This research was funded by US DOE Office of Science.

Institutional Review Board Statement: Not applicable.

Informed Consent Statement: Not applicable.

Data Availability Statement: Not applicable.

Acknowledgments: The authors thank all beamline teams and external users with whom we worked, database administrators Jeff Patton and Peter Parker, beamline scientists Changwoo Do, William T. Heller, Mark Lumsden, and all SPICE developers, all EPICS and CS-Studio developers, and all previous and current developers in the SNS Instrument Data Acquisition and Controls group. This project used resources at HFIR and SNS, DOE Office of Science User Facilities operated by ORNL. The Bio-SANS instrument is supported by the DOE Office of Biological and Environmental Research.

Conflicts of Interest: The authors declare no conflict of interest.

References

1. Heller, W.T.; Cuneo, M.; Debeer-Schmitt, L.; Do, C.; He, L.; Heroux, L.; Littrell, K.; Pingali, S.V.; Qian, S.; Stanley, C.; et al. The suite of small-angle neutron scattering instruments at oak ridge national laboratory. *J. Appl. Crystallogr.* **2018**, *51*, 242–248. [CrossRef]
2. Zolnierczuk, P.A.; Riedel, R.A. Neutron scattering experiment automation with python. In Proceedings of the 17th IEEE-NPSS Real Time Conference, Lisbon, Portugal, 24–28 May 2010; pp. 1–3. [CrossRef]
3. Lumsden, M.D.; Robertson, J.L.; Yethiraj, M. SPICE—Spectrometer and instrument control environment. *Phys. B Condens. Matter.* **2006**, *385*, 1336–1339. [CrossRef]
4. Peterson, P.F.; Campbell, S.I.; Reuter, M.A.; Taylor, R.J.; Zikovsky, J. Event-based processing of neutron scattering data. *Nucl. Instrum. Methods Phys. Res. Sect. Accel. Spectrometers Detect. Assoc. Equip.* **2015**, *803*, 24–28. [CrossRef]
5. Granroth, G.E.; An, K.; Smith, H.L.; Whitfield, P.; Neuefeind, J.C.; Lee, J.; Zhou, W.; Sedov, V.N.; Peterson, P.F.; Parizzi, A.; et al. Event-based processing of neutron scattering data at the spallation neutron source. *J. Appl. Crystallogr.* **2018**, *51*, 616–629. [CrossRef]
6. Hartman, S.M. System design towards higher availability for large distributed control systems. In Proceedings of the ICALEPCS, Grenoble, France, 10–14 October 2011; pp. 1209–1211.
7. Hartman, S.M. SNS instrument data acquisition and controls. In Proceedings of the ICALEPCS, San Francisco, CA, USA, 6–11 October 2013; pp. 755–758.
8. Geng, X.; Chen, X.H.; Kasemir, K.U. First EP- ICS/CSS based instrument control and acquisition system at ORNL. In Proceedings of the ICALEPCS, San Francisco, CA, USA, 6–11 October 2013; pp. 763–765.
9. Yao, X.; Gregory, R.; Guyotte, G.; Hartman, S.; Kasemir, K.-U.; Lionberger, C.; Pearson, M. *UX Focused Development Work During Recent ORNL Epics-Based Instrument Control System Upgrade Projects*; JACOW Publishing: Geneva, Switzerland, 2020; pp. 818–823. [CrossRef]
10. Clausen, M.R.; Gerke, C.H.; Moeller, M.; Rickens, H.R.; Hatje, J. Control system studio (CSS). In Proceedings of the ICALEPC, Knoxville, TN, USA, 15–19 October 2007; pp. 37–39.
11. Kasemir, K.U.; Pearson, M.R. CS-studio scan system parallelization. In Proceedings of the ICALEPCS, Melbourne, Australia, 17–23 October 2015; pp. 517–520. [CrossRef]
12. Kasemir, K.U.; Grodowitz, M.L. CS-studio display builder. In Proceedings of the ICALEPCS, Barcelona, Spain, 8–13 October 2017; pp. 1978–1981. [CrossRef]

13. Berry, K.D.; Bailey, K.M.; Beal, J.; Diawara, Y.; Funk, L.; Steve Hicks, J.; Jones, A.B.; Littrell, K.C.; Pingali, S.V.; Summers, P.R.; et al. Characterization of the neutron detector upgrade to the GP-SANS and bio-SANS instruments at HFIR. *Nucl. Instrum. Methods Phys. Res. Sect. Accel. Spectrometers Detect. Assoc. Equip.* **2012**, *693*, 179–185. [CrossRef]
14. De Maria Antolinos, A.; Pernot, P.; Brennich, M.E.; Kieffer, J.; Bowler, M.W.; Delageniere, S.; Ohlsson, S.; Malbet Monaco, S.; Ashton, A.; Franke, D.; et al. ISPyB for BioSAXS, the gateway to user autonomy in solution scattering experiments. *Acta Crystallogr. D Biol. Crystallogr.* **2015**, *71*, 76–85. [CrossRef] [PubMed]
15. Norman, D.; Nielsen, J. The Definition of User Experience (UX). Available online: https://www.nngroup.com/articles/definition-user-experience/ (accessed on 30 November 2020).
16. Farrell, S. UX Research Cheat Sheet. Available online: https://www.nngroup.com/articles/ux-research-cheat-sheet/ (accessed on 30 November 2020).
17. Tischler, L. Ideo's David Kelley on "Design Thinking". Available online: https://www.fastcompany.com/1139331/ideos-david-kelley-design-thinking (accessed on 30 November 2020).
18. Mildner, D.F.R.; Carpenter, J.M. Optimization of the experimental resolution for small-angle scattering. *J. Appl. Cryst.* **1984**, *17*, 249–256. [CrossRef]
19. Gerstung, H.; Elliott, C.B.; Haberman, E. Definitions of Managed Objects for Network Time Protocol Version 4 (NTPv4). 2010. Available online: https://www.hjp.at/doc/rfc/rfc5907.html (accessed on 30 November 2020).
20. Yang, Y.; Samolyuk, G.D.; Chen, T.; Poplawsky, J.D.; Lupini, A.R.; Tan, L.; Ken, L. Coupling computational thermodynamics with density-function-theory based calculations to design L12 precipitates in FeNi based alloys. *Mater. Des.* **2020**, *191*, 108592. [CrossRef]

Article

Grazing Incidence Small-Angle Neutron Scattering: Background Determination and Optimization for Soft Matter Samples

Tetyana Kyrey [1], Marina Ganeva [1], Judith Witte [2], Artem Feoktystov [1], Stefan Wellert [2] and Olaf Holderer [1,*]

[1] Forschungszentrum Jülich GmbH, Jülich Centre for Neutron Science (JCNS) at Heinz Maier-Leibnitz Zentrum (MLZ), 52425 Garching, Germany; t.kyrey@fz-juelich.de (T.K.); m.ganeva@fz-juelich.de (M.G.); a.feoktystov@fz-juelich.de (A.F.)

[2] Institute of Chemistry, Technical University Berlin, 10623 Berlin, Germany; judith.witte@campus.tu-berlin.de (J.W.); s.wellert@tu-berlin.de (S.W.)

* Correspondence: o.holderer@fz-juelich.de

Featured Application: Scattering contributions in a grazing incidence small-angle neutron scattering experiment come not only from the evanescent wave, but may also come from small-angle scattering of the transmitted beam. Simulations can be used to quantify this contribution. Optimization of the sample cell may reduce an additional background scattering from the cell.

Abstract: Grazing incidence small-angle neutron scattering (GISANS) provides access to interfacial properties, e.g., in soft matter on polymers adsorbed at a solid substrate. Simulations in the frame of the distorted wave Born approximation using the BornAgain software allow to understand and quantify the scattering pattern above and below the sample horizon, in reflection and transmission, respectively. The small-angle scattering from the interfacial layer, visible around the transmitted beam, which might contribute also on the side of the reflected beam, can be understood in this way and be included into the analysis. Background reduction by optimized sample cell design is supported by simulations, paving the way for an optimized GISANS cell.

Keywords: neutron scattering; simulation; GISANS; BornAgain Software; grazing incidence scattering

1. Introduction

In the field of soft matter, interface coatings and interface layers consisting of polymers, polymer microgels, brushes or lipid bilayers are very promising in context of practical application in medicine, smart coatings or switching devices [1–4]. However, investigation and characterisation of the interfaces, especially those buried in the sample, are very challenging and can be done by very limited numbers of experimental techniques. Neutron reflectometry allows to study the scattering length density with sub-nanometer resolution in normal direction to the interface [5], while the grazing incidence geometry gives access to the in-plane direction, i.e. to the lateral correlations within the interfacial layers. Grazing incidence small-angle neutron scattering (GISANS) allows a unique access to the interfacial regions, e.g., solid-liquid or liquid-liquid interfaces. Additional information can be obtained by combining grazing incidence neutron scattering and contrast variation via H_2O/D_2O mixing, which has been used e.g., in studies of phospholipid bilayers or diblock copolymers [6–8].

Very recently Jaksch and co-authors gave an overview of the field of grazing incidence scattering (GIS) for in-situ and operando investigations in soft matter and biophysics [9]. While GISANS is established at most neutron sources as a technique that gives access to the internal lateral structure of soft matter system at the solid or water/oil interface [9–16], grazing incidence neutron spin echo spectroscopy (GINSES) is a rather novel extension of

GISANS which allows to study the dynamics of thin polymer films, microemulsions, phospholipid membranes, etc. under similar conditions at the solid/liquid interface [7,17–20].

All these investigations have in common that only a small amount of material contributes to the scattering signal. Therefore for the most precise measurements in this field, background optimisation is of high importance in the experiment [21]. Significant care has to be taken to avoid unnecessary background contributions and to provide a sample environment with long time stability for long experimental times. Such improve of experiments can be done by combining new ways of simulation of the results and according adaption of the experimental conditions [22]. In this case, simulations of the scattering experiments may help (i) to understand the different scattering contributions to the measured signal and (ii) to interpret the experimental data.

In the first part of this paper, we present an example where GISANS experiments together with simulations help to understand the scattering signal and background contributions on the side of reflection and transmission of the GISANS cell. Since scattering under grazing incidence is treated within the distorted wave Born approximation (DWBA) [23], simulations were performed with BornAgain being a modern and easy-to-use open-source software package suitable for simulating GISANS scattering (and also the equivalent x-ray scattering GISAXS) within the DWBA [24,25]. In the second part, we propose a simple version of sample cell for GISANS experiment aiming decrease the background scattering from the cell parts. In both parts, the SANS contribution to the GISANS signal is discussed.

2. Materials and Methods

2.1. Samples of Interest

PEG-microgels based on the monomer 2-(2-methoxyethoxy)ethyl methacrylate (MEO2MA), the comonomer poly(ethylene glycol) methyl ether methacrylate (OEGMA) and the cross-linker ethylene glycol dimethacrylate (EGDMA) were synthesized by precipitation polymerization. The microgels were deposited on the solid surface by spincoating an aqueous suspensions with a microgel content of 5 wt% onto the pure silicon block. The microgels of the form p(MEO2MA-co-OEGMA) were one of our test systems we investigated in the first part of the report. Further details can be found in Ref. [11].

PNIPAM-microgels based on the monomer N-isopropylacrylamide (NIPAM) and the cross-linker N,N'-methylenebisacrylamide (BIS) were synthesised via surfactant-free precipitation polymerisation, with a cross-linker content of 0.5 mol%. Microgels were deposited onto the PEI-coated silicon blocks via spin coating for 150 s with 500 rpm from aqueous microgel dispersion. Further details can be found in Ref. [13]. PNIPAM microgels were used in the test of the new self-developed cell.

2.2. Principles of Grazing Incidence Scattering

A description of grazing incidence scattering and the distorted wave Born approximation (DWBA) can be found in the literature, e.g., in Refs. [9,23,26]. Here we just mention briefly the main characteristics of a soft matter GISANS experiment.

Typically, soft samples undergo structural changes when adsorbing to a solid substrate. For example, spherical microgel particles can be distorted and compressed when adsorbed at an interface [13]. Studying such soft matter systems consisting of polymer layers of the order of 100 nm is a challenging task due to the essentially very small scattering volume. Nevertheless, neutron reflectometry and scattering under grazing incidence allows us to obtain valuable information on the scattering length density normal to the surface as well as on lateral (in-plane) correlations at the interface, which correspond to thermal density fluctuations or spacial arrangements with a characteristic length scale [27].

GISANS scattering is normally treated in the DWBA, which takes into account multiple scattering processes at the interface [25,27,28]. According to Ref. [29] the diffuse scattered intensity (that accompanies the specular reflection) coming from lateral scattering length density fluctuations with intensity:

$$I(q) \sim \langle [F_{DWBA}(q)]^2 \rangle S(q), \qquad (1)$$

where the term $F_{DWBA}(q)$ is a DWBA form factor accounting for different scattering scenarios with multiple scattering events as described e.g., in Ref. [23].

GISANS is therefore predestined for studying the lateral inhomogeneities. In the case of soft matter samples, lateral structuring is rather blurred and lacking clear Bragg peaks or Bragg sheets, therefore a more detailed modelling of the experiment is needed.

In grazing incidence small-angle scattering (GISAS) experiment incident radiation impinges at a shallow incident angle α_i ($<1°$) and the scattered intensity is detected with a position sensitive detector as a function of scattering (α_f) and out-of-plane (ψ) angles (see Figure 1). The sample surface is the (x, y)-plane with the incident neutron beam coming along the x-axis, and the scattering plane where the detector is located is the (y, z)-plane [30].

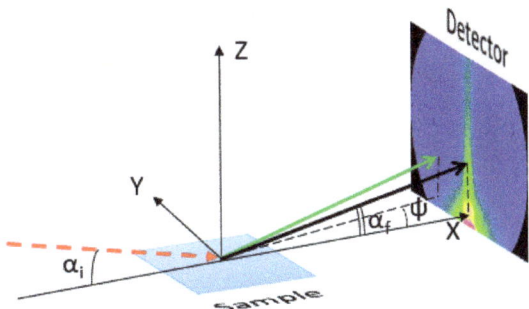

Figure 1. Scheme of grazing incidence scattering experiment. Red dashed and black solid arrows represent incoming and scattered beam, while green solid arrows corresponds to the diffuse scattering. α_i, α_f, ψ – incidence, scattering and out-of-plane angle, respectively.

At $\alpha_f = \alpha_i$ specular reflection occurs. The interesting part however for GISANS is the off-specular or diffuse scattering, where $\alpha_f \neq \alpha_i$ [31].

A characteristic part of the detector image in a GISANS experiment is the so-called Yoneda peak [32], which appears at $\alpha_f = \alpha_c$, with the critical angle of total reflection, α_c. The cut through the 2D detector pattern at the position of the Yoneda peak provides information on correlations in the lateral density distribution, for example the average inter-particle distance, the size and the shape of the nanostructures in the near surface region [26,33]. When $\alpha_i > \alpha_c$, the beam is partly reflected and partly refracted, the thin film is then fully illuminated.

If the incidence angle at which the neutron beam hits the sample surface is smaller than the critical angle of total reflection, $\alpha_i < \alpha_c$, the beam is totally reflected and a specular reflection appears on the detector. However, inside the medium an evanescent intensity distribution with the signature of a propagating wave in z-direction parallel to the surface is induced. This "evanescent wave" arises as a real part of the solution of the Schrödinger equation at $\alpha_i < \alpha_c$ (a mathematical description of the evanescent wave nature can be found elsewhere [34–36]).

The evanescent wave (EW) probes the sample only to a short distance from the probed interface (up to 100 nm) [16]. The intensity of the EW drops exponentially with the distance.

In this work, we used the software BornAgain [25] to mimic a GISANS experiment as close as possible with its 2D detector image for a given incident angle. The sample, in this case microgel particles adsorbed at the silicon interface, was modelled as core-shell truncated spheres with pre-knowledge from neutron reflectometry and AFM measurements [24]. The scattering signal was then obtained on small-angle neutron scattering instruments in grazing incidence geometry.

2.3. Grazing Incidence Small-Angle Scattering Experiment

GISANS experiments were performed on KWS-1 [37,38] and KWS-2 [39,40] small-angle neutron scattering instruments operated by Jülich Centre for Neutron Science at Maier-Leibnitz Zentrum (Garching, Germany).

Measurements on KWS-1 were performed at a sample-to-detector distance of 20 m using an unpolarised, monochromatic incident beam with a neutron wavelength of 5 Å ($\Delta\lambda/\lambda = 10\%$) under an angle of incidence (α_i) of 0.2°. Experiments on KWS-2 were conducted at sample-to-detector distances of 4 m with a neutron wavelength of 5 Å ($\Delta\lambda/\lambda = 20\%$) and under an angle of incidence (α_i) of 0.7°. This angle slightly above the critical angle was chosen to probe the structure of the entire layer of microgel particles.

In GISANS experiments samples were mounted in cells developed for grazing incidence geometry: PEG microgels in aluminium cell [41] and PNIPAM microgels in self-made quartz cell. Both cells were filled with D_2O providing high scattering contrast to the polymers and low incoherent scattering.

3. Results and Discussion

3.1. Transmission Contribution to GIS Pattern

GISANS experiments together with the corresponding simulations with BornAgain were performed on samples of PEG-microgels adsorbed at a silicon interface and are presented in Figure 2. The same GISANS data have been presented in Ref. [11], with a preliminary evaluation of the line cuts which did not cover all details of the experimental data. The angle of incidence $\alpha_i = 0.7°$ at a wavelength of 5 Å slightly above the critical angle of total reflection has been chosen to strongly illuminate the microgels.

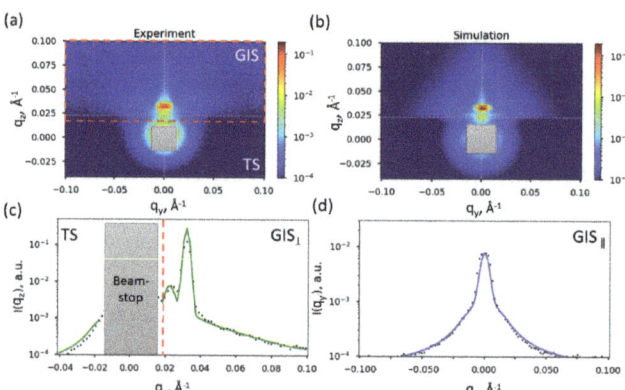

Figure 2. Experimental (**a**) and simulated with BornAgain (**b**) 2D grazing incidence small-angle neutron scattering (GISANS) scattering pattern of adsorbed p(MEO2MA-co-OEGMA) microgels. White dashed lines in (**a**,**b**) correspond to the cuts depicted in (**c**,**d**). In (**c**,**d**) the perpendicular (along q_x) and the parallel (along q_y) to the sample surface cuts are presented, dots and solid lines in (**c**,**d**) correspond to experimental and simulated data, respectively. Grey masks in (**a**–**c**) is beam-stop that blocks direct beam. The red dashed line in (**c**) represents the sample horizon.

The analysis of 2D GISANS detector images is mainly focused on the part containing the totally reflected beam above the sample horizon, whereas the part below the sample horizon, where the transmitted beam and small angle scattering still reaches the detector, is often ignored. A first analysis of the experiments on PEG-microgels at the interface was done in this way (as presented in Ref. [11]).

If the incident angle α_i was slightly above the critical angle of total reflection (α_c) as in case of PEG-microgels, transmission scattering (TS) occurred in addition to the specular and diffusive signal. Therefore, in the following analysis of the scattering 2D pattern, it was recognized that the transmitted scattering may in particular also have extended into the detector half of the GIS signal, contributing to the diffuse scattering as a background.

The transmission signal was reconstructed with BornAgain by applying the PEG-microgel model [26], and included into the GIS simulation. The scattering below the sample horizon in the TS part corresponded to the scattering of a transmission experiment at the surface layer with a slightly inclined beam. Thus, the incident angle α_i was set to 90° for the transmission scattering simulation. A direct calculation in grazing incidence geometry would be more appropriate, but is at the moment not yet implemented in the BornAgain software. Accounting for the SANS signal in this way well explained in our case the scattering contributions of the GISANS pattern and seemed to be adequate in this case. It should be noted, that such a sample rotation in virtual experiment is possible in case of an isotropic system in the probed length-scale. If structural anisotropy at the considered length scale comes into play, care has to be taken concerning details of the transmission scattering pattern.

In contrast to the highly collimated and monochromatic beam of grazing incidence small angle X-ray scattering experiments, GISANS experiments normally use a neutron beam with a much broader width and a wavelength distribution of 10–20% in order to obtain enough intensity. This leads to some peculiarities, which should not be overlooked during the data treatment and simulations.

The influence of the beam divergence on the scattering volume of the sample and on the footprint of the beam on the detector is depicted in Figure 3. The difference in scattering volume of areas 1 and 2 corresponded to the different scattering intensity on the 2D scattering pattern in the regions AB and BC. Angles are expanded in Figure 3 compared to a real experiment.

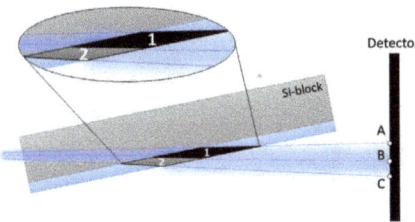

Figure 3. Schematic illustration of the transmission in grazing incidence geometry: difference in the footprints of the upper- and lower- beam parts due to the beam divergence.

The difference in the intensity distribution in area AB and BC (Figure 3) was taken into account in the following way. A matrix of the 2D scattering signal was multiplied with a vector, which provided more intense scattering in the upper part of the TS pattern (intensity ratio of 2.5:1 was taken, which depended on the scattering geometry and beam size). Simulating the lower transmission part is also a means of judging possible scattering contribution from the environment and the sample cell. If a simulated GIS pattern differs from the experimental observation additional contributions to the background should be considered.

Figure 2 shows the experimental and the simulated 2D GISANS signal including the contribution of the transmitted scattering in the direct beam direction [26]. The simulated 2D pattern as well as line cuts at $q_y = 0$ (along q_z) and at α_c (parallel to the sample surface) were in good agreement with the experimental data. There are still some remaining differences in the diffuse scattering at the borders of the upper detector part (Figure 2a,b), which might be attributed to the instrumental effects not included into the BornAgain simulation, or sample effects such as the particle size distribution, which were also not part of the simulation. The chosen beam parameters allowed an estimation of lateral correlation length, while resolution in the direction normal to the surface led to some crossing over of specular and Yoneda peak. Here it should also be noted that reproduction of profile of direct beam, such as simulation of the beam collimation, angular divergence etc., is a difficult task and is not too relevant in this case. We focus more on regions of large q, and diffuse scattering fluctuations, while the beam profile could have a stronger influence on low-q scattering or Bragg-peaks. This is still to be improved for future investigation of larger real space distances.

3.2. Test of Self-Made Cell with Reduced Background

With the previous observation that the background coming from the small-angle scattering of the sample might play a role and that also the side of the transmitted beam might influence a proper data evaluation, a new GISANS cell design with minimized material around the sample was tested. PNIPAM microgels adsorbed on Si-block were used as a test sample.

Figure 4 shows a photograph of the self-made quartz cell (a), its components (b) and the principal beam path for reflection and transmission. Basically, a quartz glass backplate was placed on a frame with boron containing glass for neutron absorption, both were directly glued together with the Si-block which contained at its surface the sample (PNIPAM-microgel particles). Small gaps on one side with syringe needles allowed us to fill the cell with water. The borated glass was used to reduce possible scattering by the glue at the interface between the glass layers. What is the optimum condition and material in this respect has to be studied in more detail in the future. D_2O was used to increase scattering contrast to the PNIPAM microgels and avoid high incoherent contribution as could be in case of H_2O.

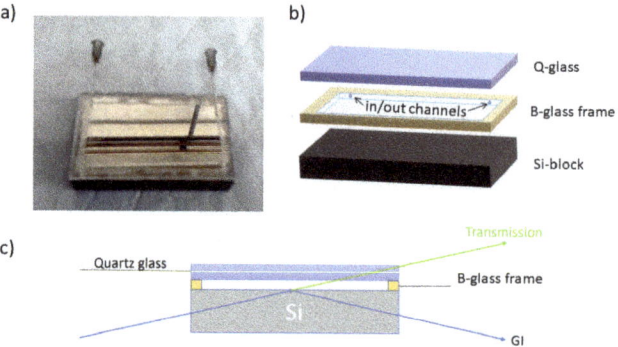

Figure 4. Cell design with minimum amount of auxiliary material. (**a**) shows a photograph of the cell, the assembly is presented in (**b**). The beam path for reflected and transmitted beam is shown in (**c**).

This cell design with its minimum surrounding opens the new possibility of measuring the dynamics of polymer systems in grazing incidence geometry by means of grazing incidence neutron spin echo spectroscopy (GINSES). Due to the instrumental setup of neutron spin echo spectrometer and very limited area for the in- and out-path for the beam usually used in an aluminium cell, currently GINSES can only be measured off-specular in q_z direction at $q_y=0$. With proposed cell design it would be possible to rotate the sample cell for a scan in the direction of q_y, giving access to lateral dynamics along the interface.

The first GISANS experiment with the proposed cell is presented in Figure 5. The left side of the detector image contained the part with the reflected beam and the GISANS scattering. On the right side of the detector (within green frame) the transmitted beam was visible. The line cut along the q_z direction at $q_y = 0$ is presented on the right side of Figure 5.

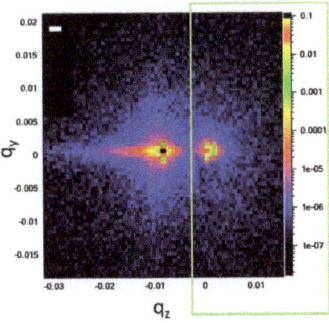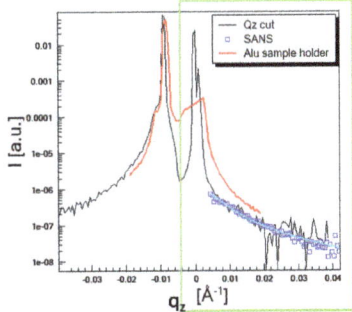

Figure 5. *Left*: Detector image of microgels at interfaces with the new cell: the GISANS part (left side of the detector) and the SANS part around the transmitted beam (within green frame). *Right*: Line cuts of the detector images along q_z at $q_y = 0$ (black line), compared to the data measured with Al cell (red line). A significantly reduced background signal is obtained. The radially averaged SANS intensity of the right side of the detector is shown as blue squares with a fit with an Ornstein–Zernike contribution.

The line cuts of the data obtained with our cell with minimal material (black line) were compared to that of the often used design with aluminum front- and back-plates with water channels for thermalization (red line). On the transmission side, additionally the radially averaged SANS intensity of this side of the detector is presented (blue dots) with a fit of the data (cyan) with an Ornstein–Zernike model for thermal fluctuations [13,42].

A significantly reduced background could be achieved with the proposed new design, which was necessary for the low-intensity GISANS experiments with biological or soft matter systems and allowed us to focus only on the scattering of the desired parts of the sample.

The self-made cell had the advantage that only a minimum amount of cell material was required, with the disadvantage that the temperature control was not possible with such a cell in the usual way done with Al cell and had to be provided by other means. For example, a thermalized air jet in the sample area, as it is used also for other "conventional" experiments, e.g., in Ref. [43] for the thermal control of an in-situ DLS setup.

4. Conclusions

The low intensity of typical samples in GISANS experiments in the field of soft matter and biology requires thorough experimental planning, data reduction and data interpretation. On the experimental side, the sample cell is a crucial part of the experiment. It has been shown that not only the side of the detector containing the reflected beam and the GISANS data, but also the backside with the transmitted beam contains valuable information and helps understanding different contributions to the signal and to the background. The small-angle scattering signal, i.e., the form factor of the monolayer of microgels present around the transmitted beam, contributed in the current experiment to a background which is not flat. The evaluation in Ref. [11] has been done without such a contribution and showed a significant deviation between measured data and optimal fitting curve with an Ornstein-Zernike model. The gain provided by simulating this SANS contribution with BornAgain provides a means of taking this into account for the interpretation of the GISANS images. Taking into account this additional background which also extends into the upper half of the detector above the horizon of the scattering plane is important if low intensities are evaluated. Soft matter GISANS experiments suffer in many cases from low intensity and rather blurred features on the detector, making this kind of analysis necessary.

The BornAgain simulation software is a very valuable tool for assessing the background contributions. This has been shown here for GISANS of microgel particles at the solid interface, and also helped in understanding background contributions for GINSES experiments (see Ref. [24]).

Testing a sample cell design with very little amount of material for GISANS showed very promising results in terms of a reduced background from the sample cell. This design would in principle also allow for new scan mode for GINSES experiments in the q_y plane to study dynamics, e.g., diffusive processes parallel to the interface.

Grazing Incidence Neutron Scattering provides unique information on the structure and also on dynamics at interfaces. It is a challenging work to study such small amounts of material as it is common for soft matter samples, but if the necessary care is taken concerning experimental background considerations and parallel simulation of the experiment, valuable information on the interface structure can be obtained.

Author Contributions: conceptualization, T.K., O.H., S.W., J.W., A.F. methodology, O.H.; software, M.G.; formal analysis, T.K.; writing—original draft preparation, O.H., T.K.; writing—review and editing, S.W. All authors have read and agreed to the published version of the manuscript.

Funding: This research was funded by Deutsche Forschungsgemeinschaft (DFG) (grant Nos. HO 5488/2-1 (OH) and WE 5066/3-1 (SW)).

Data Availability Statement: Data presented in this study can be obtained from the authors upon request.

Acknowledgments: The authors gratefully acknowledge the financial support provided by JCNS to perform the neutron scattering measurements on the KWS-1 and KWS-2 instruments at the Heinz Maier-Leibnitz Zentrum (MLZ), Garching, Germany. Fruitful discussion with Henrich Frielinghaus is acknowledged.

Conflicts of Interest: The authors declare no conflict of interest.

Abbreviations

The following abbreviations are used in this manuscript:

MDPI   Multidisciplinary Digital Publishing Institute
DOAJ   Directory of open access journals
TLA   Three letter acronym
LD   Linear dichroism

References

1. Gao, Y.; Serpe, M.J. Light-Induced Color Changes of Microgel-Based Etalons. *ACS Appl. Mater. Interfaces* **2014**, *6*, 8461–8466. [CrossRef]
2. Nolan, C.M.; Serpe, M.J.; Lyon, L.A. Pulsatile Release of Insulin from Layer-by-Layer Assembled Microgel Thin Films. *Macromol. Symp.* **2005**, *227*, 285–294. [CrossRef]
3. Uhlig, K.; Wegener, T.; He, J.; Zeiser, M.; Bookhold, J.; Dewald, I.; Godino, N.; Jaeger, M.; Hellweg, T.; Fery, A.; Duschl, C. Patterned Thermoresponsive Microgel Coatings for Noninvasive Processing of Adherent Cells. *Biomacromolecules* **2016**, *17*, 1110–1116. [CrossRef]
4. Cichosz, S.; Masek, A.; Zaborski, M. Polymer-based sensors: A review. *Polym. Test.* **2018**, *67*, 342–348. 2018.03.024. [CrossRef]
5. Majkrzak, C.F.; Berk, N.F.; Krueger, S.; Dura, J.A.; Tarek, M.; Tobias, D.; Silin, V.; Meuse, C.W.; Woodward, J.; Plant, A.L. First-principles determination of hybrid bilayer membrane structure by phase-sensitive neutron reflectometry. *Biophys. J.* **2000**, *79*, 3330–3340. [CrossRef]
6. Mangiapia, G.; Gvaramia, M.; Kuhrts, L.; Teixeira, J.; Koutsioubas, A.; Soltwedel, O.; Frielinghaus, H. Effect of benzocaine and propranolol on phospholipid-based bilayers. *Phys. Chem. Chem. Phys.* **2017**, *19*, 32057–32071. [CrossRef]
7. Jaksch, S.; Lipfert, F.; Koutsioumpas, A.; Mattauch, S.; Holderer, O.; Ivanova, O.; Frielinghaus, H.; Hertrich, S.; Fischer, S.F.; Nickel, B. Influence of ibuprofen on phospholipid membranes. *Phys. Rev. E* **2015**, *91*, 022716. [CrossRef]
8. Müller-Buschbaum, P.; Cubitt, R.; Petry, W. Nanostructured Diblock Copolymer Films: A Grazing Incidence Small-Angle Neutron Scattering Study. *Langmuir* **2003**, *19*, 7778–7782. [CrossRef]
9. Jaksch, S.; Gutberlet, T.; Müller-Buschbaum, P. Grazing incidence scattering - status and perspectives in soft matter and biophysics. *Curr. Opin. Colloid Interface Sci.* **2019**, *42*, 73–86. [CrossRef]
10. Nouhi, S.; Hellsing, M.S.; Kapaklis, V.; Rennie, A.R. Grazing-incidence small-angle neutron scattering from structures below an interface. *J. Appl. Crystallogr.* **2017**, *50*, 1066–1074. [CrossRef]

11. Kyrey, T.; Ganeva, M.; Gawlitza, K.; Witte, J.; von Klitzing, R.; Soltwedel, O.; Di, Z.; Wellert, S.; Holderer, O. Grazing incidence SANS and reflectometry combined with simulation of adsorbed microgel particles. *Phys. B Phys. Condens. Matter* **2018**, *551*, 172–178. [CrossRef]
12. Müller-Buschbaum, P. The active layer morphology of organic solar cells probed with grazing incidence scattering techniques. *Adv. Mater.* **2014**, *26*, 7692–7709. [CrossRef] [PubMed]
13. Kyrey, T.; Witte, J.; Pipich, V.; Feoktystov, A.; Koutsioubas, A.; Vezhlev, E.; Frielinghaus, H.; von Klitzing, R.; Wellert, S.; Holderer, O. Influence of the cross-linker content on adsorbed functionalised microgel coatings. *Polymer* **2019**, *169*, 29–35. [CrossRef]
14. Müller-Buschbaum, P.; Wolkenhauer, M.; Wunnicke, O.; Stamm, M.; Cubitt, R.; Petry, W. Structure formation in two-dimensionally confined diblock copolymer films. *Langmuir* **2001**, *17*, 5567–5575. [CrossRef]
15. Wellert, S.; Richter, M.; Hellweg, T.; von Klitzing, R. Responsive Microgels at Surfaces and Interfaces. *Z. Phys. Chem.* **2014**, *229*, 1225–1250. [CrossRef]
16. Müller-Buschbaum, P. GISAXS and GISANS as metrology technique for understanding the 3D morphology of block copolymer thin films. *Eur. Poly. J.* **2016**, *81*, 470–493. [CrossRef]
17. Nylander, T.; Soltwedel, O.; Ganeva, M.; Hirst, C.; Holdaway, J.; Arteta, M.Y.; Wadsa, M. Relationship between structure and fluctuations of lipid nonlamellar phases deposited at the solid-liquid interface. *J. Phys. Chem. B* **2017**, *121*, 2705–2711. [CrossRef]
18. Lipfert, F.; Frielinghaus, H.; Holderer, O.; Mattauch, S.; Monkenbusch, M.; Arend, N.; Richter, D. Polymer enrichment decelerates surfactant membranes near interfaces. *Phys. Rev. E* **2014**, *89*, 042303. [CrossRef] [PubMed]
19. Frielinghaus, H.; Kerscher, M.; Holderer, O.; Monkenbusch, M.; Richter, D. Acceleration of membrane dynamics adjacent to a wall. *Phys. Rev. E* **2012**, *85*, 041408. [CrossRef]
20. Gawlitza, K.; Ivanova, O.; Radulescu, A.; Holderer, O.; von Klitzing, R.; Wellert, S. Bulk phase and surface dynamics of PEG microgel particles. *Macromolecules* **2015**, *48*, 5807–5815. [CrossRef]
21. Hoogerheide, D.P.; Heinrich, F.; Maranville, B.B.; Majkrzak, C.F. Accurate background correction in neutron reflectometry studies of soft condensed matter films in contact with fluid reservoirs. *J. Appl. Crystallogr.* **2020**, *53*, 15–26. [CrossRef]
22. Treece, B.W.; Kienzle, P.A.; Hoogerheide, D.P.; Majkrzak, C.F.; Lösche, M.; Heinrich, F. Optimization of reflectometry experiments using information theory. *J. Appl. Crystallogr.* **2019**, *52*, 47–59. [CrossRef] [PubMed]
23. Lazzari, R. IsGISAXS: a program for grazing-incidence small-angle X-ray scattering analysis of supported islands. *J. Appl. Crystallogr.* **2002**, *35*, 406–421. [CrossRef]
24. Kyrey, T.; Ganeva, M.; Witte, J.; von Klitzing, R.; Wellert, S.; Holderer, O. Understanding near-surface polymer dynamics by a combination of grazing-incidence neutron scattering and virtual experiments. *J. Appl. Crystallogr.* **2021**, *54*. [CrossRef]
25. Pospelov, G.; Van Herck, W.; Burle, J.; Carmona Loaiza, J.M.; Durniak, C.; Fisher, J.M.; Ganeva, M.; Yurov, D.; Wuttke, J. BornAgain: software for simulating and fitting grazing-incidence small-angle scattering. *J. Appl. Crystallogr.* **2020**, *53*, 262–276. [CrossRef] [PubMed]
26. Kyrey, T. Internal Structure and Dynamics of PNIPAM Based Microgels in Bulk and Adsorbed State at Different Internal Crosslinker Distributions. Ph.D. Thesis, Technische Universität, Darmstadt, Germany, 2019.
27. Müller-Buschbaum, P. Grazing incidence small-angle neutron scattering: challenges and possibilities. *Poly. J.* **2013**, *45*, 34–42. [CrossRef]
28. Korolkov, D. *Structural Analysis of Diblock Copolymer Nanotemplates Using Grazing Incidence Scattering*; Forschungszentrum Jülich: Jülich, Germany, 2008.
29. Santoro, G.; Yu, S. Grazing incidence Small Angle X-ray Scattering as a Tool for In-Situ Time-Resolved Studies; In *X-ray Scattering*; Ares, A.E., Ed.; INTECH: London, UK, 2017; Chapter 2, pp. 29–60.
30. Müller-Buschbaum, P.; Gutmann, P.; Stamm, M.; Cubitt, R.; Cunis, S.; von Krosigk, G.; Gehrke, G.; Petry, W. Dewetting of thin polymer blend films: Examined with GISAS. *Phys. B* **2000**, *283*, 53–59. [CrossRef]
31. Lauter-Pasyuk, V. Neutron grazing incidence techniques for nano-science. *Collect. SFN* **2007**, *7*, 221–240. [CrossRef]
32. Yoneda, Y. Anomalous surface reflection of X rays. *Phys. Rev.* **1963**, *131*, 2010–2013. [CrossRef]
33. Schwartzkopf, M.; Roth, S.V. Investigating polymer-metal interfaces by grazing incidence small-angle X-ray scattering from gradients to real-time studies. *Nanomaterials* **2016**, *6*, 239. [CrossRef] [PubMed]
34. Milosevic, M. On the nature of the evanescent wave. *Appl. Spectrosc.* **2013**, *67*, 126–131. [CrossRef]
35. Knoll, W. Polymer thin films and interfaces characterized with evanescent light. *Die Makromolekulare Chemie* **1991**, *192*, 2827–2856. [CrossRef]
36. Dosch, H. *Critical Phenomena at Surfaces and Interfaces*; Springer Tracts in Modern Physics: Berlin/Heidelberg, Germany, 1992.
37. Feoktystov, A.V.; Frielinghaus, H.; Di, Z.; Jaksch, S.; Kleines, H.; Ioffe, A.; Richter, D. KWS-1 high–resolution small–angle neutron scattering instrument at JCNS: current state. *Appl. Crystallogr.* **2015**, *48*, 61–70. [CrossRef]
38. Zentrum, H.M.L. KWS-1: Small–angle scattering diffractometer. *J. Large-Scale Res. Facil.* **2015**, *1*, A28.
39. Zentrum, H.M.L. KWS-2: Small angle scattering diffractometer. *J. Large-Scale Res. Facil.* **2015**, *1*, A29. [CrossRef]
40. Radulescu, A.; Pipich, V.; Frielinghaus, H.; Appavou, M.-S. KWS-2, the high intensity/wide Q-range small-angle neutron diffractometer for soft-matter and biology at FRM II. *J. Phys. Conf. Ser.* **2012**, *351*, 012026. [CrossRef]
41. Steitz, R.; Gutberlet, T.; Hauss, T.; Klo, B.; Krastev, R.; Schemmel, S.; Simonsen, A.C.; Findenegg, G.H. Nanobubbles and their precursor layer at the interface of water against a hydrophobic substrate. *Langmuir* **2003**, *19*, 2409–2418. [CrossRef]

42. Shibayama, M. Small–angle neutron scattering on polymer gels: phase behavior, inhomogeneities and deformation mechanisms. *Polym. J.* **2011**, *43*, 18–34. [CrossRef]
43. Balacescu, L.; Vögl, F.; Staringer, S.; Ossovyi, V.; Brandl, G.; Lumma, N.; Feilbach, H.; Holderer, O.; Pasini, S.; Stadler, A.; et al. In situ dynamic light scattering complementing neutron spin echo measurements on protein samples. *J. Surf. Investig. X-ray Synchrotron Neutron Tech.* **2020**, *14*, S185–S189. [CrossRef]

Communication

Light Scattering and Absorption Complementarities to Neutron Scattering: In Situ FTIR and DLS Techniques at the High-Intensity and Extended Q-Range SANS Diffractometer KWS-2

Livia Balacescu [1,2], Georg Brandl [2], Fumitoshi Kaneko [3], Tobias Erich Schrader [2] and Aurel Radulescu [2,*]

[1] Physikalisches Institut (IA), Rheinisch-Westfälische Technische Hochschule (RWTH), Otto-Blumenthal Str., 52074 Aachen, Germany; liviab1992@gmail.com
[2] Forschungszentrum Jülich GmbH, Jülich Centre for Neutron Science (JCNS) at Heinz Maier-Leibnitz Zentrum (MLZ), 85747 Garching, Germany; g.brandl@fz-juelich.de (G.B.); t.schrader@fz-juelich.de (T.E.S.)
[3] Graduate School of Science, Osaka University, 1-1 Machikaneyama, Toyonaka, Osaka 560-0043, Japan; toshi@chem.sci.osaka-u.ac.jp
* Correspondence: a.radulescu@fz-juelich.de; Tel.: +49-89-289-10712

Citation: Balacescu, L.; Brandl, G.; Kaneko, F.; Schrader, T.E.; Radulescu, A. Light Scattering and Absorption Complementarities to Neutron Scattering: In Situ FTIR and DLS Techniques at the High-Intensity and Extended Q-Range SANS Diffractometer KWS-2. *Appl. Sci.* **2021**, *11*, 5135. https://doi.org/10.3390/app11115135

Academic Editor: Antonino Pietropaolo

Received: 30 April 2021
Accepted: 29 May 2021
Published: 31 May 2021

Publisher's Note: MDPI stays neutral with regard to jurisdictional claims in published maps and institutional affiliations.

Copyright: © 2021 by the authors. Licensee MDPI, Basel, Switzerland. This article is an open access article distributed under the terms and conditions of the Creative Commons Attribution (CC BY) license (https://creativecommons.org/licenses/by/4.0/).

Abstract: Understanding soft and biological materials requires global knowledge of their microstructural features from elementary units at the nm scale up to larger complex aggregates in the micrometer range. Such a wide range of scale can be explored using the KWS-2 small-angle neutron (SANS) diffractometer. Additional information obtained by in situ complementary techniques sometimes supports the SANS analysis of systems undergoing structural modifications under external stimuli or which are stable only for short times. Observations at the local molecular level structure and conformation assists with an unambiguous interpretation of the SANS data using appropriate structural models, while monitoring of the sample condition during the SANS investigation ensures the sample stability and desired composition and chemical conditions. Thus, we equipped the KWS-2 with complementary light absorption and scattering capabilities: Fourier transform infrared (FTIR) spectroscopy can now be performed simultaneously with standard and time-resolved SANS, while in situ dynamic light scattering (DLS) became available for routine experiments, which enables the observation of either changes in the sample composition, due to sedimentation effects, or in size of morphologies, due to aggregation processes. The performance of each setup is demonstrated here using systems representative of those typically investigated on this beamline and benchmarked to studies performed offline.

Keywords: SANS; FTIR; DLS; semi-crystalline polymers; proteins in buffer solution

1. Introduction

Small-angle neutron scattering (SANS) is a powerful technique that yields important low-resolution structural information on macromolecular systems. The small-angle neutron diffractometer, KWS-2 [1], operated by the Jülich Centre for Neutron Science (JCNS) at the Heinz Maier-Leibnitz Zentrum (MLZ), Garching, Germany, is dedicated to the investigation of mesoscopic multi-scale structures and structural changes in soft condensed matter and biophysical systems. Following demands from the user community, it was recently considerably upgraded to boost its performance with respect to the intensity on the sample, instrumental resolution, counting rate capabilities, and maximum Q-range ($Q = (4\pi/\lambda)\sin\theta$ is the momentum transfer, where λ is the neutron wavelength and 2θ is the scattering angle). The instrument, with high-intensity, extended Q-range and tunable resolution, is optimized for the study of mesoscopic structures and structural changes due to rapid kinetics.

The ability to observe chemical and physical processes simultaneously with small-angle scattering (SAS) is highly desirable for a deeper understanding of biological and soft matter systems. This is particularly crucial for dynamic systems, to ensure sample stability, purity, chemical conditions, to reduce ambiguities in data reduction, and for when it is not possible to perform further measurements or characterization on intermediate states ex situ. Given the high brilliance of synchrotron radiation, simultaneous analyses by X-ray and differential scanning calorimetry (DSC) or Raman spectroscopy were developed quite early at the Synchrotron Radiation Source in Daresbury, U.K. [2,3], and used successfully since then [4,5]. More recently, in situ size-exclusion chromatography (SEC) with SAXS [6] was developed for the first time at the Advanced Photon Source of Argonne National Laboratory (ANL, Lemont, IL, USA). This brought advantages in the field of protein scattering, in which samples very often typically show poor stability or aggregation tendency.

As unique probes, neutrons interact differently with the ^{1}H and ^{2}H (deuterium, D) hydrogen isotopes: the large difference in the coherent scattering length density between H and D represents the basis of the contrast variation and contrast matching methods. As most of the soft matter and biological samples consist of hydrocarbon systems, the H/D substitution offers the possibility to vary the coherent scattering length density of a compound over a broad range. With this technique, selected constituents in a complex multi-component system can be labeled by isotope exchange. Depending on the created contrast—the squared difference between the scattering length density of one component and that of the other components or the environment medium—selected components or regions within a complex soft matter or biophysical morphology can be made visible or invisible in the scattering experiment without the chemical alteration of the system.

Owing to technological developments in high-intensity specialized SANS instruments such as detectors and neutron guides [7,8], exciting developments in the nature and number of characterization techniques that can be performed in situ on large-scale SANS instruments started to emerge. For example, the first implementation of an in situ SEC-SANS system was reported with a D22 instrument at the Institute Laue-Langevin (ILL, Grenoble, France) [9], while first results obtained with a simultaneous SANS and FTIR measuring system were collected with the KWS-2 instrument [10]. In situ DLS setups have also been developed to complement SANS at several neutron beamlines so far, for example a SANS-I instrument at the SINQ spallation source (Paul-Scherrer Institute, PSI, Villigen, Switzerland) [11], D11 instrument at the Institute Laue-Langevin (ILL, Grenoble, France) [12,13], and LOQ and ZOOM instruments at the ISIS Pulsed Neutron and Muon Source (Rutherford Appleton Laboratory, Didcot, UK) [14].

Therefore, the simultaneous combination of light scattering or absorption methods with SANS (or WANS) represents very powerful experimental approaches to be used in the microstructural characterization of soft and biological systems.

In this paper we communicate recent upgrades to the sample environments available using the KWS-2, which is now equipped to perform in situ FTIR and dynamic light scattering (DLS). All of which can be performed simultaneously with standard or real-time SANS and enable in-beam monitoring of the sample quality or provide complementary information to neutrons.

2. Materials and Methods

Small-angle neutron scattering (SANS) experiments were performed using the KWS-2 instrument at the JCNS at MLZ in Garching, Germany [1]. The measurements were typically performed using sample-to-detector distances of 1.5, 8, and 20 m, and neutron λ of 5 and 10 Å ($\Delta\lambda/\lambda = 10\%$), to achieve a maximum dynamic q-range of 1×10^{-3}–0.7 Å^{-1}. Scattering data was collected in static and real-time modes. Each raw scattering data set was corrected for detector sensitivity, electronics background, and empty cell contribution and converted to scattering cross-section data using instrument software QtiKWS [15]. Data was then converted to the absolute scale (cm^{-1}) through reference to the scattering from a secondary standard sample (Plexiglass).

Hydrogenated polyethylene glycol dimethyl ether (hPEGDME, M_W = 20,000 g mol^{-1} with a polydispersity of Đ = 1.01) that was synthesized in house [16] was dissolved in deuterated tetrahydrofuran (dTHF, Sigma-Aldrich Chemie GmbH, Munich, Germany) by weighing the components in appropriate amounts to ensure a polymer volume fraction (vol/vol%) ϕ_{pol} = 2% in the final solution. The solution was first heated to 50 °C under stirring, to achieve a homogeneous distribution of polymer chains in solution and then transferred into a sample container with ZnSe windows (Laser components GmbH, Olching, Germany), which is appropriate for simultaneous FTIR and SANS measurements. The sealed sample container with a beam path length of 1 mm was installed in a Peltier-controlled holder of own production that can enable the variation of temperature on the sample between 5 °C and 130 °C in a controlled way. To determine the crystallization temperature of PEGDME from the dTHF solution in decreasing temperature, the system was investigated by differential scanning calorimetry (DSC) using a DSC131 EVO Calorimeter (SETARAM Instrumentation, KEP Technologies EMEA, Kirchheim unter Teck, Germany) prior to the simultaneous FTIR and SANS analysis.

Uniaxially oriented amorphous deuterated syndiotactic polystyrene (d-sPS) films, about 50 μm thick, were prepared by quenching a melt of d-sPS in an ice water bath, drawing the melt-quenched film four times at 373 K, and clipping well-oriented portions from the drawn films [17]. d-sPS co-crystal films containing protonated toluene as guest (d-sPS/hTol) were obtained by exposing the oriented amorphous films to a vapor of hTol (Sigma-Aldrich Chemie GmbH, Munich, Germany). The d-sPS/hTol co-crystalline films were transferred into the sample container with ZnSe windows for simultaneous FTIR and SANS analysis at room temperature.

In situ FTIR investigations were carried out during the SANS measurements with a JASCO VIR-200 spectrometer (JASCO Deutschland GmbH, Pfungstadt, Germany) installed at the sample position of the KWS-2 in an appropriate geometry that enabled a simultaneous irradiation of the sample using IR and neutron coaxial beams. Polarized IR beams were obtained by using a precision automated PIKE polarizer KRS-5 (PIKE Technologies, Madison, WI, USA).

For the in situ DLS measurements, a 15 mg/mL solution of apomyoglobin (apoMb) molten globule was prepared. ApoMb was obtained from horse heart myoglobin (Sigma-Aldrich Chemie GmbH, Munich, Germany) by the butanone method (as performed in [18]) and then refolded by dialysis in 20 mM NaH_2PO_4/Na_2HPO_4 (Sigma-Aldrich Chemie GmbH, Munich, Germany) pH 7 buffer. Before storage in the freezer at −20 °C, the solution was lyophilized. To replace the exchangeable protons by deuterium, the freeze-dried apoMb powder was dissolved in heavy water (D_2O, 99.9% D, Sigma-Aldrich Chemie GmbH, Munich, Germany), incubated for 1 day, and lyophilized again. To obtain the molten globule state of apoMb the powder was dissolved in D_2O and centrifuged to remove the large aggregates. In the supernate solution of concentration 2 mg/mL and pD 6, DCl 0.1 M (Sigma-Aldrich Chemie GmbH, Munich, Germany) was added until the pH value was 3.6 (monitored by pH meter Methrom), corresponding to a pD value of 4. The buffer exchanged protein solution was centrifuged (Heraeus Instruments) to the final concentration of 10 mg/mL using Vivaspin 3000 MWCO concentration units (Sartorius, Göttingen, Germany). The protein concentration was determined in a 0.1 mm thick cell (Hellma, Germany), using the molar extinction coefficient ($\epsilon 280$ nm = 13,980 M^{-1} cm^{-1}) which was calculated from the amino acid sequence [19].

All samples were prepared in deuterium solvents or using deuterated polymer films to provide low incoherent neutron background and enable appropriate scattering contrasts for neutrons.

3. Results

In the following sections we will present results and discussion of the proof-of-principle tests on the in situ FTIR and DLS setups performed using the KWS-2 simultaneously with the SANS characterization of samples.

3.1. In Situ FTIR

FTIR allows for the determination of the chemical and physical structure of material via interpretation of the spectrum of characteristic bands that is produced by the excitation vibrations of molecular bonds following absorption of IR photons in the material [20]. FTIR is particularly powerful in the analysis of the polymeric systems under different environmental conditions. The chemical identity of the chains is achieved by detecting the chemical groups present in the polymer, the monomer sequences, and the stereoregularity. Additionally, one can observe the local conformation of individual chains as well as conformational change when macroscopic fields such as temperature, pressure, and relative humidity are applied to the sample. Due to these advantages, a combination of FTIR and SANS represents an outstanding investigation approach for understanding the detailed mechanism of structure formation and structural changes in multiphase polymeric systems such as semi-crystalline polymers, copolymers, or polymer composites: the conformational change of the molecules that make up such a complex system can be monitored by FTIR, while the hierarchy of morphologies occurring in different phases can be qualitatively and quantitatively characterized by SANS with contrast variation. Polymeric materials are very sensitive to slight variations of external conditions; therefore, it is more useful to apply the combination of these techniques in a simultaneous approach, when in situ or time-resolved analyses of rapid structural changes or irreversible processes can be performed.

We previously conducted pioneering work in developing a simultaneous SANS and FTIR spectroscopy measuring system using the KWS-2 diffractometer in 2014. Using this approach we studied the structure of a d-sPS co-crystalline film with various guest molecules [10]: higher-order crystalline structures were characterized by SANS while the local molecular configuration and molecular packing were characterized by FTIR. We demonstrated that the FTIR data were useful for interpreting and analyzing SANS profiles when larger molecules are included in the sPS films [21]. To extend the measurable Q-range for wide-angle neutron scattering (WANS), we used this approach on the time-of-flight SANS instrument TAIKAN installed at the Material and Life Science Experimental Facility (MLF), Japan Proton Accelerator Research Complex (J-PARC, Tokai, Japan) [22,23]. The old measurement system tested with the KWS-2 consisted of a portable FTIR spectrometer Perkin-Elmer Spectrum Two (PerkinElmer, Rodgau, Germany) and an optical system made up of six mirrors. For an easier optical path adjustment, higher light utilization efficiency, and proper use of the available space on the sample stage of the KWS-2, and for the installation of a device for the measurement of polarization-dependent infrared absorption and a more complex sample environment, the optical system including the FTIR spectrometer was redesigned, using the experience accumulated in [22]. A new FTIR spectrometer (JASCO VIR-200, JASCO Deutschland GmbH, Pfungstadt, Germany) is currently in regular use for conducting FTIR measurements in transmission geometry, with the IR beam going through the sample coaxially with the neutron beam (Figure 1): the two optical ports in the same lateral face, the outlet port for emitting parallel light and the inlet port for the incoming light, enable a great flexibility in designing the optical system and accommodating special sample environments (Figure 1b,c). At the same time, this geometry is paving the way for the installation of an in situ ATR-FTIR option that can be used simultaneously with SANS in the near future for an optimal investigation of liquid samples. In the current geometry, only the two Al-coated quartz plates (LASER COMPONENTS GmbH, Olching, Germany) are left (elements 3 and 4 in the scheme of Figure 1a).

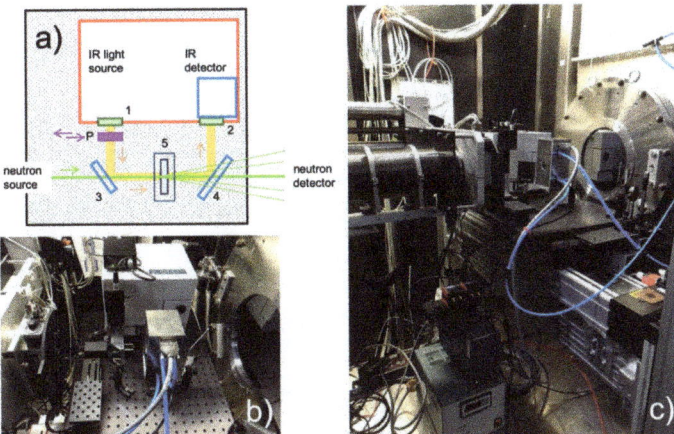

Figure 1. Schematic presentation (**a**) and detailed (**b**) and general (**c**) views of the new operational setup for in situ FTIR analysis simultaneous with the SANS investigation using the KWS-2. The numbers indicate the main components, which can be identified in the photos on the (**b**,**c**) panels: the outlet and inlet beam ports (1,2), the windows (3,4), and the new ancillary equipment (5) for the control of the temperature on the sample (in-house designed and produced Peltier-based thermostated cuvette holder), while P indicates the IR polarizer.

Figure 1b,c shows photos of the new setup for simultaneous FTIR and SANS analysis using the KWS-2: the components indicated by numbers in the scheme in Figure 1a are identified in the photographs, which show the complete experimental arrangement at the sample position of the KWS-2 between the "collimation nose" that holds the sample aperture on the left side of photos, and the entrance window of the evacuated SANS detector tank, which is visible on the right side of the images. The new FTIR setup is entirely built on a breadboard (Thorlabs GmbH, Bergkirchen, Germany) which is equipped with "zero-point" fixation systems (AMF GmbH, Fellbach, Germany) for pneumatic clamping (activated with pressurized air) for an easy and precise installation on the sample stage of the KWS-2.

The control of the FTIR spectrometer, IR polarizer position (in beam and out of beam) and orientation were implemented in the NICOS instrument control software of the KWS-2 [24]. This is mainly meant for time-resolved investigations when FTIR measurements with the selected time acquisition can be carried out simultaneously with SANS measurements on samples in different external fields or composition conditions. The start of the FTIR acquisition is triggered by the detector start of the SANS instrument and runs until the SANS acquisition time is consumed. All relevant parameters for SANS: wavelength λ, collimation length L_C, detection length L_D, beam-size on sample, etc. [15], sample field parameters (temperature), and the FTIR setup parameters (polarizer position and orientation, etc.) can be combined in a single complex experimental routine. For conventional SANS measurements, standard FTIR acquisitions can be carried out in parallel using the operation menu of the FTIR spectrometer. A precise synchronization of the results obtained with the two methods is possible by using the SANS data acquisition in list mode [8], which enables the assignment of the SANS pattern collected at a specific time to the corresponding FTIR spectrum.

Proof-of-principle measurements were carried out by simultaneous use of FTIR in transmission geometry and SANS on soft matter systems that are relevant for the research topics developed using the KWS-2: (i) a hPEGDME solution in dTHF (ϕ_{pol} = 2%) at 30 °C and 12 °C and (ii) a d-sPS film in co-crystalline δ-form with hTol as guest molecules in the cavities between the sPS helices.

3.1.1. FTIR Sample Holder and Neutron Background

Figure 2 shows a technical drawing of the sample cells designed for the liquid and film samples for the simultaneous FTIR and SANS setup using the KWS-2. The cell is equipped with two ZnSe windows, each with a thickness of 1 mm, which enable a beam path of either 1 mm or 2 mm. A shorter beam path can be obtained by placing an additional ZnSe window with a thickness of 0.5 mm and an appropriate diameter inside of the holder. The chosen geometry allows for an easy filling and a tight closing of cells using the series of lateral screws, appropriate O-rings, and Teflon screws for the top openings used for filling in the samples. Tests of the holders have shown that they remain tight up to a temperature of at least 120 °C.

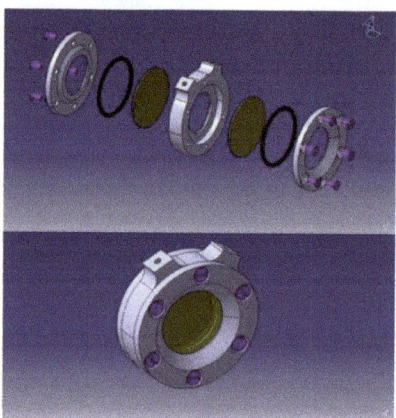

Figure 2. Technical drawing of the new sample cuvette equipped with ZnSe windows, which is used for the simultaneous FTIR and SANS investigation of samples using the KWS-2. The liquid samples can be filled in by using the top openings, while the black O-rings, the series of screws (violet), and Teflon screws for the top openings provide the cuvette tightness. Cuvettes with 1 mm and 2 mm sample thickness are provided; the sample thickness can be further decreased to 0.5 mm or 0.2 mm by using additional ZnSe windows of appropriate thickness and diameter that can be placed inside of the cuvette, between the yellow windows shown here.

The background yielded by the set of Al-coated mirrors and ZnSe windows in SANS is low (Figure 3) and, despite the weak Bragg reflexes from the ZnSe material, which can be observed in the high-Q scattering conditions (the small spots indicated by arrows in Figure 3c), very good background subtraction for SANS data of high quality can be achieved.

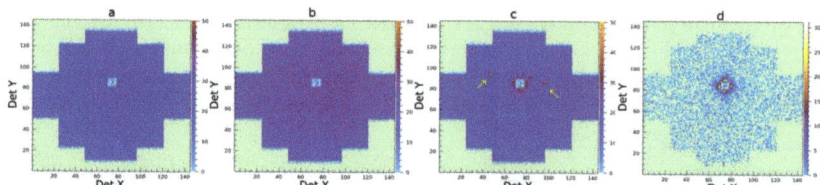

Figure 3. The two-dimensional SANS patterns corrected for detector sensitivity collected at L_D = 1.5 m from: empty beam (**a**), when the two Al-coated quartz windows (items 3 and 4 in Figure 1a) are added (**b**), and when the two ZnSe windows are added (**c**); data from the same setup as in (**c**) but collected at L_D = 20 m are shown in (**d**). The yellow arrows indicate the Bragg reflections from the ZnSe material.

3.1.2. Simultaneous FTIR and SANS on PEGDME in Solution

Unlike crystallization from bulk or melt, a variety of processes, such as the formation of polymer and solvent compounds, a physical gelation, or a liquid–liquid phase separation, may take place when polymers crystallize from solution, depending on the solvent type and polymer concentration and molecular weight [25].

PEGDME is a semi-crystalline polymer, soluble in water and compatible with most organic solvents. When PEGDME crystallizes from solution, crystallites grow as aggregates of small layers of lamellae with branching and splitting. The morphology of this hierarchical organization of structures over a wide length scale, between nm and µm, can be resolved by SANS. In the crystalline state, PEGDME takes a regular helical conformation, which can be regarded approximately as a uniform (7/2) helix [26,27]. When taking a long helical structure, PEGDME chains exhibit some sharp conformational regularity IR bands, such as the 1345 cm^{-1} A$_2$ band [28,29]. Thus, the combination between SANS and in situ FTIR is a powerful method for the study of the kinetics and morphology of polymer crystallization from solution.

In Figure 4 we report the results of the DSC scan taken under nitrogen atmosphere between 50 °C and 10 °C from the solution of the PEGDME in dTHF with a cooling rate of 5 °C/min. The polymer crystallization occurs at about T$_C$ = 26–27 °C, with two peaks observed in the heat flow curve, an effect similar to that observed for the bulk crystallization of PEG [30]. Therefore, in order to investigate the change in conformation of PEGDME chains and the formation of polymer crystalline morphology at mesoscopic scale when passing through the T$_C$, we performed the simultaneous FTIR and SANS analysis of the hPEGDME in dTHF solution at 30 °C, above the T$_C$, followed by a rapid cooling to 12 °C, below the T$_C$.

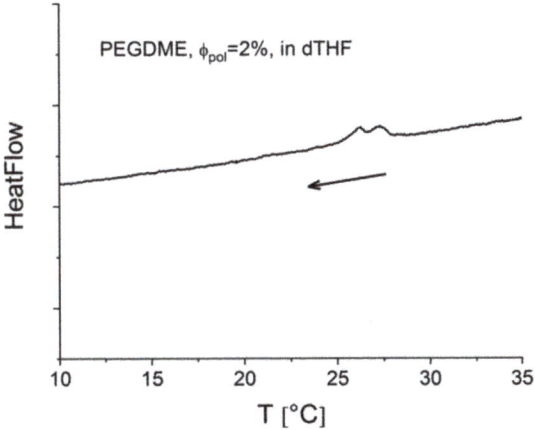

Figure 4. The thermal behavior of hPEGDME (M$_W$ = 20,000 g mol^{-1}) in solution of dTHF (ϕ_{pol} = 2%) during the crystallization process at a rate of 5 °C/min.

Figure 5 shows a selected range of the infrared spectra from the hPEGDME in dTHF at 30 °C (blue line) and at 12 °C (with the cyan and red lines indicating the initial and final stages, and the other lines in between depicting the time evolution of the spectrum at this temperature) in parallel with the spectrum of dTHF taken at 30 °C. Upon reaching the temperature of 12 °C, infrared spectra from the sample were collected every 30 s within a time interval of 35 min, to follow the kinetics for isothermal crystallization of polymer in solution. The time-resolved analysis clearly shows the evolution and transformation of the band at 1352 cm^{-1}, which is characteristic of amorphous polymer structure (single coil in solution) and well visible in the initial stage at 12 °C, into the bands at 1345 cm^{-1} and 1364 cm^{-1}, which are characteristic of the helical conformation of polymer chain in the crystalline state and well visible in the final stage at 12 °C. All other bands, corresponding to either the amorphous conformation in the initial stage or the helical conformation in the final stage, correspond to frequencies reported for the PEG system [31,32].

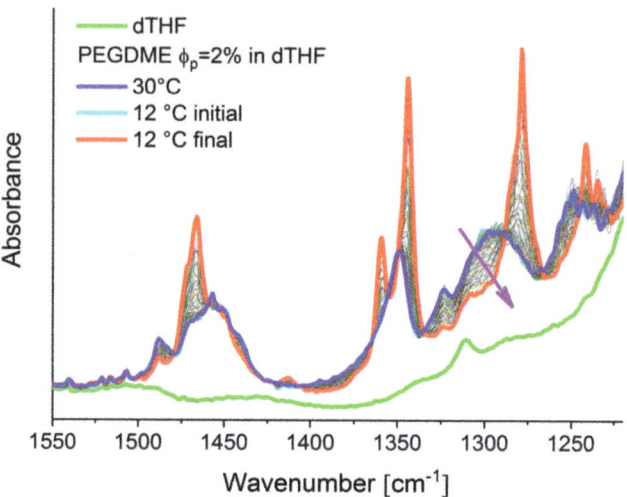

Figure 5. FTIR spectral change observed during the crystallization of hPEGDME (M_W = 20,000 g mol^{-1}) from solution in dTHF in decreasing temperature from 30 °C to 12 °C. The arrow indicates the evolution of the FTIR spectra collected between the initial and final states at 12 °C with a time step of 30 s.

The polymer morphology in solution at 30 °C and 12 °C in the initial and final stages of crystallization was investigated by SANS in parallel with the FTIR analysis. Figure 6 reports the corrected and calibrated SANS results collected simultaneously with the acquisition of the infrared spectra. The power law exponents (Q^{-p}) specific for different polymer morphologies are given. Hence, swollen single coils ($p = 5/3$), one-dimensional morphologies or Kuhn segments ($p = 1$), lamellar ($p = 2$), and large 3D morphologies with sharp interfaces ($p = 4$) can be directly identified and characterized [33]. The model curves of the semi-flexible single coil and the crystalline lamellar stacks morphologies described later are also depicted.

At 30 °C all polymers are dissolved as semi-flexible coils in solution, as shown by the single-chain form factor features identified in the scattering profiles: the plateau towards low Q yields information about the volume fraction and molecular weight while the bending down of the intensity is characteristic of the Guinier regime, from which information about the radius of gyration of the polymer coil is obtained. The $Q^{-5/3}$ power law regime towards high Q is indicative of excluded volume interactions between the chain segments (with $-5/3$—Flory's exponent). A transition to Q^{-1} behavior is observed at higher Q, which indicates that short stiff (linear) polymer segments with a persistence length $l_P = b/2$, with b the Kuhn segment, exist at a local length scale [34]. The small-angle scattering pattern at 30 °C was fitted using the unified Beaucage model [35] for multiple structural levels in hierarchical morphologies. The approach described the experimental data well and delivered the sizes involved in each structural level: the R_g = 37 Å of the polymer coil and the persistence length l_P = 7.5 Å. Thus, both FTIR and SANS results are in good agreement regarding the polymer conformation at 30 °C as semi-flexible coil in solution.

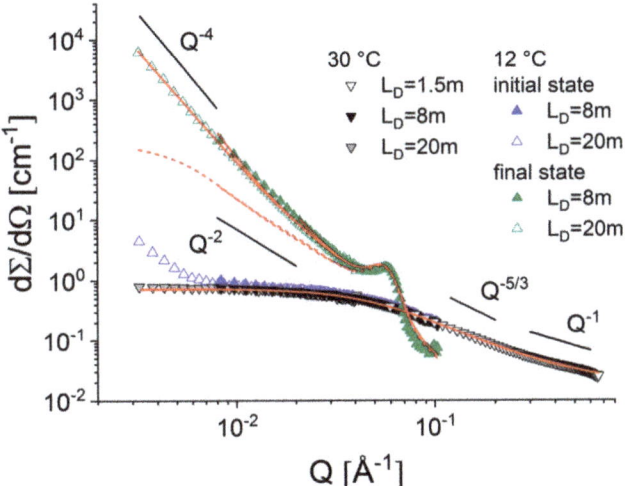

Figure 6. SANS cross sections from solution of hPEGDME (M_W = 20,000 g mol^{-1}) in dTHF (ϕ_{pol} = 2%) measured at different temperatures, during the FTIR analysis (Figure 7). Data collected at different L_D and temperatures are differentiated by symbols and colors as shown in the legend. The solid lines indicate the power law behavior in different Q ranges, whereas the red curves represent the model description of the experimental data, as discussed in the text. The dotted curve represents the scattering pattern from the lamellar stacks, as resulted from the global fit of the experimental data collected in the final state at 12 °C (green symbols).

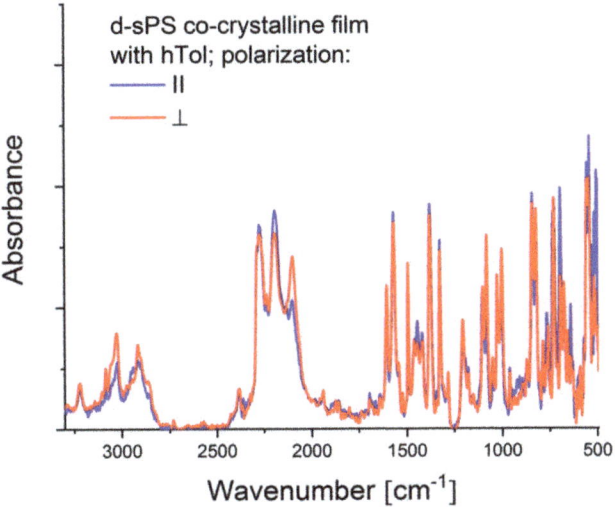

Figure 7. Cont.

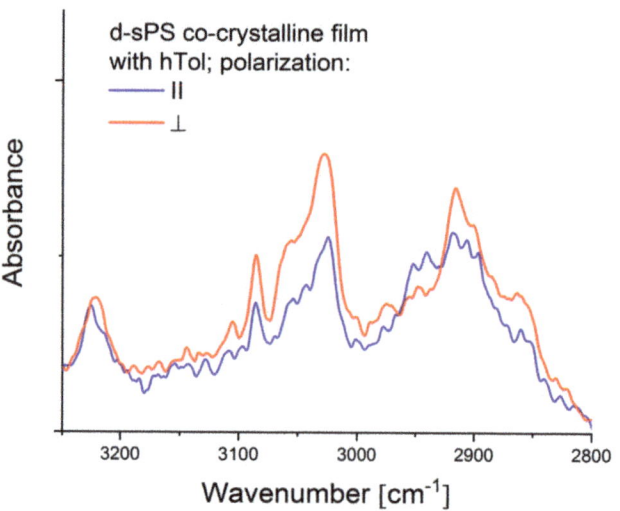

Figure 7. FTIR spectra measured with polarized IR beam parallel (blue line) and perpendicular (red line) to the uniaxially deformation direction of the d-sPS co-crystalline film with hTol: whole spectra (**top**) and detail (**bottom**) are shown.

At 12 °C, due to methodical reasons, we used the typical procedure for carrying out experiments over an extended Q-range by combining different L_D's rather than the time-resolved measurements on a selected narrow Q-range. Thus, the raw data were collected for one minute at two detection distances, L_D = 8 m and 20 m (always in this order), with the detector movement from one position to another taking 10 min. Two sets of measurements were performed at 12 °C, starting at the time when the set temperature was attained by the Peltier-based variable temperature cuvette holder, which corresponds to the acquisition start of the cyan IR spectrum in Figure 5.

A decrease in the temperature resulted in the formation of polymer crystalline morphology that led to an increase of the scattering level towards low Q compared with the single chain scattering feature, as observed in the SANS results corresponding to the initial state at 12 °C (blue symbols in Figure 6). As observed in the FTIR spectra, a coexistence between an amorphous conformation of PEGDME, on one hand, and a slow increase in number density of crystallites by nucleation and growth, on the other hand, is observed, which explains the SANS pattern.

The scattering data collected in the end state at 12 °C indicate the formation at mesoscale of a morphology made of crystalline lamellae alternating with amorphous interlamellar regions, as denoted by the peak-like feature which indicates an interlamellar domain spacing of $D^* = 2\pi/Q^*$: where Q^* represents the peak position, and the steep increase of intensity towards low Q, due to association of the lamellar stacks in bundles and branches, which led to the formation of a large scale spherulite morphology.

The scattering features characteristic of such morphology appear at much lower Q values than the range covered in pinhole mode with the KWS-2 [1] and in the current experiment only the high-Q asymptotic behavior of the scattering from the polymer spherulites could be observed (green symbols, Q^{-4} power law behavior). A fit with the model combining the form factor of individual lamellar slabs with the paracrystalline structure factor describing the stacking effects over distances larger than the lamellar thickness [33] successfully described the experimental SANS data and delivered the interlamellar distance D^* = 110 Å and its large smearing σ_{D^*} = 30 Å. A power law term with the exponent $p = 4$, as discussed above, was considered for a proper description of the data at low Q, which far exceeds the scattering from the polymer lamellae (depicted as a red dotted curve in Figure 6, based on the fitted parameters). A value for the lateral extension of lamellae (assumed as discs) much larger than their thickness was assumed as fixed parameter (R_L = 500 Å), as long as this size level cannot be observed towards low Q, while a fixed lamellar thickness d = 45 Å [17] was considered in the fitting procedure, since no scattering data were collected at high Q at 12 °C. The lateral extension of the lamellae does not affect the fitting of the D* and $σ_{D*}$ parameters, as long as it is supposed to be much larger than the lateral size of the lamellar stacks. A more detailed model interpretation of the SANS data in terms of "forward scattering" and scattering length density parameters for different regions of the complex morphology [33] is beyond the goal of this report. The SANS results indicate that crystalline lamellar stacks, which form larger objects (spherulites), are the dominating morphology in the sample and are in very good agreement with the FTIR observations, which indicate that PEGDME chains are mainly in helical conformation, thus crystalline state, in these conditions.

3.1.3. Polarized FTIR Measurements

In Figure 7 we report polarized FTIR spectra measured on a d-sPS/dTol co-crystalline uniaxially deformed film with the polarized IR radiation parallel (blue line) and perpendicular (red line) to the drawing direction of the film. The FTIR analysis was done in this case independently to the SANS investigation and was meant to prove the performance of the IR polarizer using well-known samples [21]. SANS measurements on similar systems and guest exchange process were reported elsewhere [36]. The d-sPS/hTol-oriented film exhibits clear polarization in the 3100–2800 cm^{-1} region, which suggests that toluene guest molecules are kept oriented in the crystalline region of the d-sPS, as they are located between the oriented helices of the d-sPS host system. The hTol provides IR bands due to benzene ring C–H stretch modes in 3100–3000 cm^{-1}, while its methyl group shows two bands due to antisymmetric stretch modes at about 2945 and 2920 cm^{-1} and a band due to symmetric stretch mode at about 2860 cm^{-1}.

3.2. In Situ Dynamic Light Scattering

DLS, also known as photon-correlation spectroscopy (PCS) or quasi-elastic light scattering spectroscopy (QELSS), is based on the temporal analysis of the intensity fluctuations of the scattered light caused by the Brownian motion of particles (protein molecules, aggregates, polymer particles, etc.) in solution. The velocity of the Brownian motion is defined by a property known as the translational diffusion coefficient (D). D is related to the average hydrodynamic radius of the particles (R_H) through the Stokes–Einstein equation [37]:

$$D = \frac{k_B T}{6\pi\eta R_H}$$

where k_B is the Boltzmann constant, T is the temperature, and η is the viscosity of the medium. The radius obtained by this technique is the radius of a sphere that has the same diffusion coefficient as the particle. If several particles are present in the solution, they can be easily distinguished if they are at least one order of magnitude different in size. This is particularly useful for samples, which show a tendency to aggregate over time, such as biological samples, which are sensitive to slight changes of temperature, pH of the solvent, radiation damage, etc. Since SANS measurements offering structural information on the nanometer scale often require integration times of hours, monitoring the sample composition on the minute timescale with DLS provides an additional insight into the aggregation state of the sample. This is why in situ DLS has been developed and used at many SANS beamlines so far [11–14].

A DLS measurement requires a highly monochromatic light beam, for example, produced by a laser illuminating an area of particles within a solution. The refractive index jump between those particles and the surrounding solution causes the light to be scattered. The particle size (and magnitude of Brownian motion) will affect the rate at which the intensity of this scattered light fluctuates. A single photon counting detector converts the scattered photons into standardized electric pulses. Those are fed into a digital autocorrelator which sorts the photons into time bins with respect to the time (t) when a first photon arrived at the detector and time ($t + \delta t$), when the next photon arrived on the detector. Thereby it constructs a correlation function ($g^{II}(t)$). For larger particles, the scattered light changes slowly and the correlation persists for a long time. In contrast, if the particles are small and move rapidly then the correlation function decays more quickly.

3.2.1. DLS Setup

Similar to the setup developed by Heigl et al. [12], the DLS setup using the KWS-2 is built modularly, consisting of five major components: light source, optics defining the scattering geometry, detector, correlator, and computer. The optics defining the scattering geometry are placed at the sample position on the instrument, whereas the other components are mobile connected with optical fibers (Figure 8). The light from a He–Ne laser (20 mW, 632.8 nm) is coupled to a fiber (Schäfter + Kirchhoff GmbH, Hamburg, Germany) and brought to the neutron instrument. A collimator (Schäfter + Kirchhoff GmbH, Hamburg, Germany) and a mirror focus the light from the fiber into the center of the quartz cell, which was previously aligned in the center of the neutron beam. The scattered light under the angle ϑ (in the present case $\vartheta = 120°$) is then guided by a mirror and collimated into a second fiber. The optics are mounted using standard optic holders (Thorlabs) and placed on a platform anchored on the vacuum tank of the neutron detector. This concept allows the use of a variety of sample changers with the existing optics. For alignment purposes, one can mark a spot on the quartz cuvette in the sample holder, where the laser hits the cell surface. After disconnecting the laser fiber and reconnecting it to the collimator for the scattered light, one can see and align the light path for the scattered light to the same sample spot marked in the quartz cell. This alignment is often good enough to obtain a first correlation curve which can then be optimized to a better intercept using nanoparticle solutions as highly scattering test samples. This alignment concept relies on the fact that both the laser light and the scattered light are guided through a single mode optical fiber, a specialty of this realization of in situ DLS.

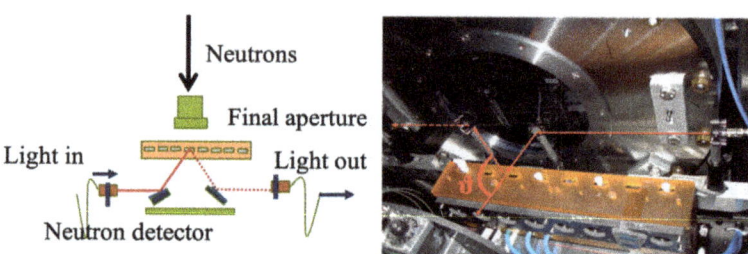

Figure 8. Left: Schematic view of the setup for simultaneous DLS and SANS analysis of samples using the KWS-2. **Right**: view of the setup on the sample stage of the neutron SANS diffractometer, between the sample aperture of the collimation system and the entrance window of the evacuated detector tank. Light beam trajectories to and from the sample are indicated in red. The sample cuvettes are placed in the locations of a Peltier-based multi-position variable temperature holder (Quantum Northwest Inc., Liberty Lake, WA, USA) that can move each sample in neutron/light beams in a controlled way.

After passing the mirror, the collimator, and the single mode fiber, the scattered photons arrive at an APD detector (Excelitas Technologies Corp. SPCM-AQRH-FC, Waltham, MA, USA). The recorded electrical pulses are autocorrelated with an ALV-7004 Multiple Tau Digital Correlator (ALV-GmbH, Langen, Germany). A computer displays the measured intensities and the autocorrelation functions. The correlation function and the recorded intensities are saved for further analysis. To protect the detection components, a filter wheel with neutral density filters is placed before coupling the laser light into the fiber.

Based on the photon counts detected for each sample, the wheel adjusts to the optimum filter for each measurement.

A DLS measurement begins shortly after the start of the neutron measurement, which has been triggered by the instrument control software NICOS. Light scattering data are recorded in parallel and saved on a separate computer with a unique identifier containing the run number of the neutron measurement. The correlator determines the normalized intensity–intensity (second-order) auto-correlation function $g^{II}(\tau)$. Data processing is performed post-experiment, by obtaining the electrical field first-order correlation function $g^{I}(\tau)$ from $g^{II}(\tau)$, according to the Siegert relation. $g^{I}(\tau)$ is interpreted using a triple-exponential fit, corresponding to a triple-modal distribution [38]. For samples with higher polydispersity, a CONTIN fit routine can be used [39].

3.2.2. Proof of Concept: Investigation of Silica Nanospheres

Silica nanospheres (20 nm, NanoXact™, nanoComposix, San Diego, CA, USA) were investigated as a proof of concept. The water-based solution with a reported concentration of 5 mg mL^{-1} was used as received. According to the producer, these nanospheres are sensitive to lower temperatures, leading to aggregation. Their diameter is 21.4 ± 2.7 nm, as determined by JEOL 1010 transmission electron microscope gravimetric analysis. In terms of absorption characteristics, their optical density at the 632.8 nm wavelength of the DLS-laser is lower than 0.005, meaning the light absorption by the silica nanosphere at this wavelength is negligible. The SANS data shown in Figure 9 confirms the silica nanosphere radius (R_{ns} = 11.9 nm with polydispersity σ = 12%) and is not altered by the formation of any aggregates. On the other hand, the DLS investigations performed in parallel reveal that there are also larger structures in the solution of the silica nanospheres. The intensity of the first-order correlation function $g^{I}(\tau)$ can be attributed as follows: 18% monomer (with a signal at 0.48 ms, corresponding to an average hydrodynamic radius of 13.3 nm) and 63% to aggregates (average R_H of 185 nm). Less than 15% of the intensity can be attributed to larger-sized aggregates (R_H ~2 µm) or can be assumed to occur because of noise. Given that the scattering intensity strongly depends on the particle's volume, the percentage corresponds to 98% monomer and 2% aggregates. The reason for the presence of aggregates can be assumed as a thermal mistreatment of the sample due to temperature variation.

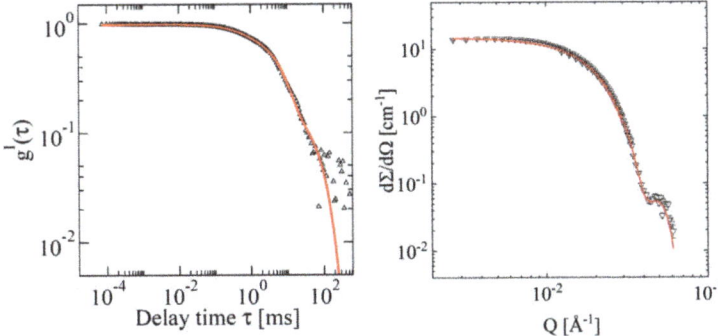

Figure 9. **Left**: the electrical field first-order correlation function $g^{I}(\tau)$ obtained from the DLS measurement of the solution of silica nanospheres (with nominal diameter of 20 nm). The red line is a triple-exponential fit, identifying two decay times: 0.48 ms and 6.64 ms with considerable amplitudes and a third one with a comparably small amplitude which will be attributed to noise here. **Right**: the SANS cross section of the same solution (open symbols). The red line is the data fit based on the spherical form factor model (R_{ns} = 11.9 nm) with polydispersity in size (σ = 12%) and instrument resolution considered [15]. Although the SANS curve does not signal the presence of aggregates in the solutions, the DLS does.

3.2.3. A Typical Sample: Apomyoglobin

Apomyoglobin in the molten globule state was investigated by SANS and in situ DLS. Because the protein is not in its native-alike state and the sample concentration is significantly higher than in the living organism, this system is prone to aggregation. Even slight changes in temperature and pH of the environment can lead to sudden aggregation. To ensure that the sample does not change during the SANS experiment, in situ DLS was employed. Every 5th minute during the hours-long SANS experiment, a DLS measurement was performed.

The first and the last measurement are shown in Figure 10 and one can see that the sample did not change. The first-order correlation function was evaluated using the CONTIN algorithm [39] and the particle-size distribution according to the scattering intensity can be seen in Figure 11. This confirms that the molten globule apomyoglobin solution did not form more aggregates during the SANS experiment and the amount of aggregates present already at the start of the SANS experiment could be determined. Beyond the R_H~2 nm monomers, larger structures of R_H ~100 nm are present in the solution. Signals of particles of $R_H > 10$ μm which are attributed to noise can also be identified. Nevertheless, based on the fact that the scattering intensity is strongly dependent on the particle's volume, the largest number of particles in the solution is represented by monomers and the second largest by the R_H ~100 nm aggregates.

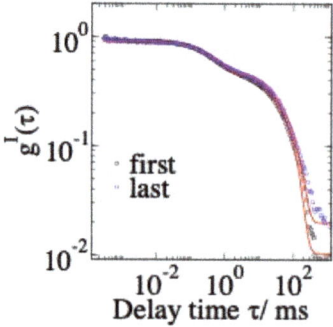

Figure 10. The electrical field first-order correlation functions of the first and the last DLS measurements performed on the apomyoglobin sample during the neutron beam time. Although larger structures were permanently present in the sample, the two curves nearly coincide. This means that the aggregation state of the sample did not change during the long integration time needed for the SANS measurement. The red lines mark the fit to the data using the CONTIN algorithm.

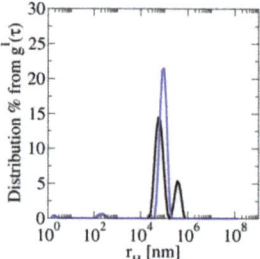 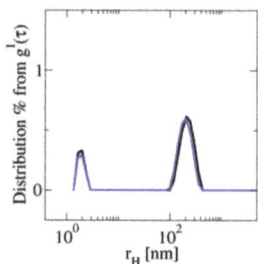

Figure 11. Left: The derived size distribution obtained from the measurements presented in Figure 12. **Right**: close-up of the smaller particles in the solution; besides the monomers with R_H ~2 nm, larger structures of R_H ~110 nm are present in the solution. The black line corresponds to the distributions obtained from the first DLS measurement, whereas the blue line indicates the results of the last measurement.

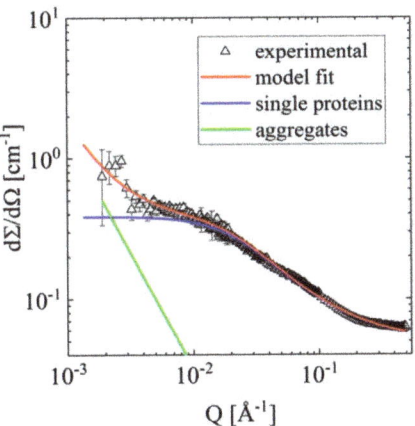

Figure 12. SANS cross sections of the 15 mg/mL apomyoglobin molten globule solution. The blue line marks the simulation for the monomer scattering intensity based on the polymer with excluded volume model (R_g = 2.5 nm, ν = 0.46). The green line is the simulated scattering curve for a larger structure. The red line is obtained by adding the green and blue lines and describes the data well.

Figure 12 shows the corresponding SANS curve to the in situ DLS data discussed above. The scattering profile of the monomeric protein apomyoglobin is represented by the blue line and was determined based on the polymer with excluded volume model [40] and previous knowledge about the protein: excluded volume parameter ν = 0.46 and R_g = 2.5 nm [41].

The green line shows the scattering contribution from the aggregates with R_H = 150 nm (which can be roughly approximated to R_g = 116 nm, according to the ratio $R_g/R_H = (3/5)^{0.5}$ [42]). Since for such large morphologies the Guinier regime occurs at lower Q values than the Q-range of the current SANS investigation, this contribution mostly resembles the power law regime of the Beaucage model with the Porod exponent p = 5/3 [35]. The red line is obtained by adding the blue and green lines, with an appropriate power law scaling factor at low Q, to describe the data well. Using this additional information from the in situ DLS measurement, one can now interpret the SANS curve better.

4. Conclusions

We reported here the concept and proof-of-principle tests of the in situ Fourier transform infrared (FTIR) spectroscopy and dynamic light scattering (DLS) analysis using the high intensity and extended Q-range SANS diffractometer KWS-2. Simultaneous approaches using light absorption and scattering with standard or real time SANS are now possible with the KWS-2, which enable in-beam control of the sample quality or provide complementary structural and composition information to neutrons. The performance of each setup was demonstrated using systems representative of those typically investigated on this beamline and benchmarked to studies performed offline.

These developments have been greatly assisted by the short exposure times provided by the high neutron flux of the FRM II reactor, the increased maneuverability allowed at the sample position of the instrument, and the flexibility of the instrument control software, NICOS, which is developed in-house at Garching.

Author Contributions: Conceptualization, L.B., T.E.S., A.R.; methodology, F.K., T.E.S., A.R.; software, G.B.; validation, L.B., F.K., T.E.S., A.R.; investigation, L.B., T.E.S., A.R.; data curation, L.B., T.E.S., A.R.; writing—original draft preparation, L.B., F.K., T.E.S., A.R.; writing—review and editing, L.B., T.E.S., A.R. All authors have read and agreed to the published version of the manuscript.

Funding: Not applicable.

Institutional Review Board Statement: Not applicable.

Informed Consent Statement: Not applicable.

Data Availability Statement: The data presented in this study are available on request from the corresponding author.

Acknowledgments: We are thankful to Simon Staringer, Vladimir Ossovyi, Thomas Kohnke, and Andreas Nebel (JCNS) for their technical support during installation and commissioning of the in situ FTIR and DLS setups with the KWS-2 SANS diffractometer. Design and realization of the Peltier-based variable temperature cuvette holder by Vladimir Ossovyi (JCNS) is specially acknowledged. The help from Maria-Maddalena Schiavone with the DSC measurements is gratefully acknowledged. A. R. is thankful to Hiroki Iwase (CROSS Neutron Science and Technology Center, Tokai, Japan) for help with the concept drawings of the sample cuvettes for the FTIR and SANS setup. This project was partially financially supported by the BMBF projects 05K16PA1 and 05K19PA3. T. E. S. wishes to thank Stefan Rustler, Simon Lechlmayr, and Raimund Heigl for their pioneering work on the in situ DLS setup.

Conflicts of Interest: The authors declare no conflict of interest.

References

1. Radulescu, A.; Pipich, V.; Frielinghaus, H.; Appavou, M.-S. KWS-2, the high intensity/wide Q-range small-angle neutron diffractometer for soft-matter and biology at FRM II. *J. Phys. Conf. Ser.* **2012**, *351*, 012026. [CrossRef]
2. Bras, W.; Derbyshire, G.E.; Devine, A.; Clark, S.M.; Cooke, J.; Komanschek, B.E.; Ryan, A.J. The Combination of Thermal Analysis and Time-Resolved X-ray Techniques: A Powerful Method for Materials Characterization. *J. Appl. Crystallogr.* **1995**, *28*, 26–32. [CrossRef]
3. Bryant, G.K.; Gleeson, H.F.; Ryan, A.J.; Fairclough, J.P.A.; Bogg, D.; Goossens, J.G.P.; Bras, W. Raman spectroscopy combined with small angle X-ray scattering and wide angle X-ray scattering as a tool for the study of phase transitions in polymers. *Rev. Sci. Instrum.* **1998**, *69*, 2114. [CrossRef]
4. Tashiro, K.; Sasaki, S. Structural changes in the ordering process of polymers as studied by an organized combination of the various measurement techniques. *Prog. Polym. Sci.* **2003**, *28*, 451–519. [CrossRef]
5. Hirose, R.; Yoshioka, T.; Yamamoto, H.; Reddy, K.R.; Tahara, D.; Hamada, K.; Tashiro, K. In-house simultaneous collection of small-angle X-ray scattering, wide-angle X-ray diffraction and Raman scattering data from polymeric materials. *J. Appl. Crystallogr.* **2014**, *47*, 922–930. [CrossRef]
6. Mathew, E.; Mirza, A.; Menhart, N. Liquid-chromatography-coupled SAXS for accurate sizing of aggregating proteins. *J. Synchrotron Radiat.* **2004**, *11*, 314–318. [CrossRef]
7. Radulescu, A.; Pipich, V.; Ioffe, A. Quality assessment of neutron delivery system for small-angle neutron scattering diffractometers of the Jülich Centre for Neutron Science at the FRM II. *Nucl. Instrum. Methods Phys. Res. A* **2012**, *689*, 1–6. [CrossRef]
8. Houston, J.E.; Brandl, G.; Drochner, M.; Kemmerling, G.; Engels, R.; Papagiannopoulos, A.; Sarter, M.; Stadler, A.; Radulescu, A. The high-intensity option of the SANS diffractometer KWS-2 at JCNS—Characterization and performance of the new multi-megahertz detection system. *J. Appl. Crystallogr.* **2018**, *51*, 323–336. [CrossRef]
9. Jordan, A.; Jacques, M.; Merrick, C.; Devos, J.; Forsyth, T.; Porcar, L.; Martel, A. SEC-SANS: Size exclusion chromatography combined in-situ with small-angle neutron scattering. *J. Appl. Crystallogr.* **2016**, *49*, 2015–2020. [CrossRef]
10. Kaneko, F.; Seto, N.; Sato, S.; Radulescu, A.; Schiavone, M.M.; Allgaier, J.; Ute, K. Development of a Simultaneous SANS/FTIR Measuring System. *Chem. Lett.* **2015**, *44*, 497–499. [CrossRef]
11. Kohlbrecher, J.; Bollhalder, A.; Vavrin, R. A high pressure cell for small angle neutron scattering up to 500MPa in combination with light scattering to investigate liquid samples. *Rev. Sci. Instrum.* **2007**, *78*, 125101. [CrossRef] [PubMed]
12. Heigl, R.J.; Longo, M.; Stellbrink, J.; Radulescu, A.; Schweins, R.; Schrader, T.E. Crossover from a Linear to a Branched Growth Regime in the Crystallization of Lysozyme. *Crystallogr. Growth Des.* **2018**, *18*, 1483–1494. [CrossRef]
13. Nawroth, T.; Buch, P.; Buch, K.; Langguth, P.; Schweins, R. Liposome Formation from Bile Salt–Lipid Micelles in the Digestion and Drug Delivery Model FaSSIFmod Estimated by Combined Time-Resolved Neutron and Dynamic Light Scattering. *Mol. Pharm.* **2011**, *8*, 2162–2172. [CrossRef] [PubMed]
14. Nigro, V.; Angelini, R.; King, S.; Franco, S.; Buratti, E.; Bomboi, F.; Mahmoudi, N.; Corvasce, F.; Scaccia, R.; Church, A.; et al. Apparatus for simultaneous dynamic light scattering–small angle neutron scattering investigations of dynamics and structure in soft matter. *Rev. Sci. Instr.* **2021**, *92*, 023907. [CrossRef]
15. Radulescu, A.; Szekely, N.K.; Appavou, M.S.; Pipich, V.; Kohnke, T.; Ossovyi, V.; Staringer, S.; Schneider, G.J.; Amann, M.; Zhang-Haagen, B.; et al. Studying Soft-matter and Biological Systems over a Wide Length-scale from Nanometer and Micrometer Sizes at the Small-angle Neutron Diffractometer KWS-2. *J. Vis. Exp.* **2016**, *118*, e54639. [CrossRef]
16. Hövelmann, C.H.; Gooßen, S.; Allgaier, J. Scale-Up Procedure for the Efficient Synthesis of Highly Pure Cyclic Poly(ethylene glycol). *Macromolecules* **2017**, *50*, 4169–4179. [CrossRef]

17. Kaneko, F.; Radulescu, A.; Ute, K. Time-resolved SANS studies on guest exchange processes in cocrystals of syndiotactic polystyrene. *Polymer* **2013**, *54*, 3145–3149. [CrossRef]
18. Stadler, A.M.; Koza, M.M.; Fitter, J. Determination of conformational entropy of fully and partially folded conformations of holo- and apomyoglobin. *J. Phys. Chem. B* **2015**, *119*, 72. [CrossRef] [PubMed]
19. Gasteiger, E. Expasy: The proteomics server for in-depth protein knowledge and analysis. *Nucl. Acids Res.* **2003**, *31*, 3784. [CrossRef]
20. Larkin, P.J. *Infrared and Raman Spectroscopy. Principles and Spectral Interpretation*; Elsevier: Amsterdam, The Netherlands, 2011.
21. Kaneko, F.; Seto, N.; Sato, S.; Radulescu, A.; Schiavone, M.M.; Allgaier, J.; Ute, K. Simultaneous small-angle neutron scattering and Fourier transform infrared spectroscopic measurements on cocrystals of syndiotactic polystyrene with polyethylene glycol dimethyl ethers. *J. Appl. Crystallogr.* **2016**, *49*, 1420–1427. [CrossRef]
22. Kaneko, F.; Kawaguchi, T.; Radulescu, A.; Iwase, H.; Morikawa, T.; Takata, S.; Nishiura, M.; Hou, Z. A new simultaneous measurement system of wide Q-range small angle neutron scattering combined with polarized Fourier transform infrared spectroscopy. *Rev. Sci. Instrum.* **2019**, *90*, 093906. [CrossRef]
23. Kaneko, F.; Radulescu, A.; Iwase, H.; Takata, S.; Nishiura, M.; Hou, Z. Application of Simultaneous Measurement System Combining Wide Q-Range Small-Angle Neutron Scattering and Polarized Fourier Transform Infrared Spectroscopy: Cocrystal of Syndiotactic Polystyrene with Methyl Benzoate. *JPS Conf. Proc.* **2021**, *33*, 011076.
24. Brandl, G.; Felder, C.; Pedersen, B.; Faulhaber, E.; Lenz, A.; Krüger, J. NICOS–The Instrument Control solution at the MLZ. In Proceedings of the 10th International Workshop on Personal Computers and Particle Accelerator Controls, Karlsruhe, Germany, 14–17 October 2014.
25. Sasaki, T.; Miyazaki, A.; Sugiura, S.; Okada, K. Crystallization of Poly(ethylene oxide) from Solutions of Different Solvents. *Polym. J.* **2002**, *34*, 794–800. [CrossRef]
26. Takahashi, Y.; Sumita, I.; Tadokoro, H. Structural studies of polyethers. IX. Planar zigzag modification of Poly(ethylene oxide). *J. Polym. Sci. A Polym. Phys.* **1973**, *11*, 2113–2122. [CrossRef]
27. Kobayashi, M.; Kitagawa, K. Microstructure of poly (ethylene oxide) gels dispersed in various organic solvents. *Macromol. Symp.* **1997**, *114*, 291–296. [CrossRef]
28. Yoshihara, T.; Tadokoro, H.; Murahashi, S.J. Normal Vibrations of the Polymer Molecules of Helical Conformation. IV. Polyethylene Oxide and Polyethylene-d_4 Oxide. *Chem. Phys.* **1964**, *41*, 2902–2911. [CrossRef]
29. Matsuura, H.; Miyazawa, T.J. Vibrational analysis of molten poly (ethylene glycol). *Polym. Sci. A Polym. Phys.* **1969**, *7*, 1735–1744. [CrossRef]
30. Gines, J.M.; Arias, M.J.; Rabasco, A.M.; Novak, C.; Ruiz-Conde, A.; Sanchez-Soto, P.J. Thermal characterization of polyethylene glycols applied in the pharmaceutical thchnology using differential scanning calorimetry and hot stage microscopy. *J. Thermal Anal.* **1996**, *46*, 291–304. [CrossRef]
31. Harder, P.; Grunze, M.; Dahint, R.; Whitesides, G.M.; Laibinis, P.E.J. Molecular Conformation in Oligo(ethylene glycol)-Terminated Self-Assembled Monolayers on Gold and Silver Surfaces Determines Their Ability To Resist Protein Adsorption. *Phys. Chem. B* **1998**, *102*, 426–435. [CrossRef]
32. Miyazawa, T.; Fukushima, K.; Ideguchi, Y.J. Molecular vibrations and structure of high polymers. III. Polarized infrared spectra, normal vibrations, and helical conformation of polyethylene glycol. *Chem. Phys.* **1962**, *37*, 2764–2776. [CrossRef]
33. Radulescu, A.; Fetters, L.J.; Richter, D. Polymer driven wax crystal control using partially crystalline polymeric materials. *Adv. Polym. Sci.* **2008**, *210*, 1–100.
34. Ramachandran, R.; Beaucage, G.; Kulkarni, A.S.; McFaddin, D.; Merrick-Mack, J.; Galiatsatos, V. Persistence Length of Short-Chain Branched Polyethylene. *Macromolecules* **2008**, *41*, 9802–9806. [CrossRef]
35. Hammouda, B. Analysis of the Beaucage model. *J. Appl. Crystallogr.* **2010**, *43*, 1474–1478. [CrossRef]
36. Kaneko, F.; Radulescu, A.; Ute, K. Time-resolved small-angle neutron scattering study on guest-exchange processes in co-crystals of syndiotactic polystyrene. *J. Appl. Crystallogr.* **2014**, *47*, 6–13. [CrossRef]
37. Cruickshank, M.C. The Stokes-Einstein law for diffusion in solution. *Proc. R. Soc. Lond. A.* **1924**, *106*, 724–749.
38. Hassan, P.; Kulshreshtha, S. Modification to the cumulant analysis of polydispersity in quasielastic light scattering data. *J. Colloid Interface Sci.* **2006**, *300*, 744–748. [CrossRef] [PubMed]
39. Provencher, S. Contin: A general purpose constrained regularization program for inverting noisy linear algebraic and integral equations. *Comput. Phys. Commun.* **1982**, *27*, 229–242. [CrossRef]
40. Petrescu, A.J.; Receveur, V.; Calmettes, P.; Durand, D.; Desmadril, M.; Roux, B.; Smith, J.C. Small-angle neutron scattering by a strongly denatured protein: Analysis using random polymer theory. *Biophys. J.* **1997**, *72*, 335–342. [CrossRef]
41. Balacescu, L.; Schrader, T.E.; Radulescu, A.; Zolnierczuk, P.; Holderer, O.; Pasini, S.; Fitter, J.; Stadler, A.M. Transition between protein-like and polymer-like dynamic behavior: Internal friction in unfolded apomyoglobin depends on denaturing conditions. *Sci. Rep.* **2020**, *10*, 1570–1580. [CrossRef]
42. Kok, C.M.; Rudin, A. Relationship between the hydrodynamic radius and the radius of gyration of a polymer solution. *Makromol. Chem. Rapid Commun.* **1981**, *2*, 655–659. [CrossRef]

Article

Simultaneous SAXS/SANS Method at D22 of ILL: Instrument Upgrade

Ezzeldin Metwalli [1], Klaus Götz [1,2,3], Tobias Zech [1,2,3], Christian Bär [1], Isabel Schuldes [1], Anne Martel [4], Lionel Porcar [4] and Tobias Unruh [1,2,3],*

1. Institute for Crystallography and Structural Physics, Friedrich-Alexander-Universität Erlangen-Nürnberg, Staudtstrasse 3, 91058 Erlangen, Germany; ezzeldin.ali@fau.de (E.M.); klaus.klagoe.goetz@fau.de (K.G.); tobias.zech@fau.de (T.Z.); christian.baer@fau.de (C.B.); isabel.schuldes@fau.de (I.S.)
2. Center for Nanoanalysis and Electron Microscopy (CENEM), Cauerstrasse 3, 91058 Erlangen, Germany
3. Interdisciplinary Center for Nanostructured Films (IZNF), Cauerstrasse 3, 91058 Erlangen, Germany
4. Institut Laue-Langevin, 71 Avenue des Martyrs, 38042 Grenoble, France; martela@ill.fr (A.M.); porcar@ill.fr (L.P.)
* Correspondence: tobias.unruh@fau.de; Tel.: +49-9131-85-25189

Abstract: A customized portable SAXS instrument has recently been constructed, installed, and tested at the D22 SANS instrument at ILL. Technical characteristics of this newly established plug-and-play SAXS system have recently been reported (*J. Appl. Cryst.* 2020, 53, 722). An optimized lead shielding arrangement on the SAXS system and a double energy threshold X-ray detector have been further implemented to substantially suppress the unavoidable high-energy gamma radiation background on the X-ray detector. The performance of the upgraded SAXS instrument has been examined systematically by determining background suppression factors (SFs) at various experimental conditions, including different neutron beam collimation lengths and X-ray sample-to-detector distances (SDD_{X-ray}). Improved signal-to-noise ratio SAXS data enables combined SAXS and SANS measurements for all possible experimental conditions at the D22 instrument. Both SAXS and SANS data from the same sample volume can be fitted simultaneously using a common structural model, allowing unambiguous interpretation of the scattering data. Importantly, advanced in situ/real time investigations are possible, where both the SAXS and the SANS data can reveal time-resolved complementary nanoscale structural information.

Keywords: small angle X-ray scattering; small angle neutron scattering; SAS; SANS; SAXS; nanomaterials

Citation: Metwalli, E.; Götz, K.; Zech, T.; Bär, C.; Schuldes, I.; Martel, A.; Porcar, L.; Unruh, T. Simultaneous SAXS/SANS Method at D22 of ILL: Instrument Upgrade. *Appl. Sci.* 2021, 11, 5925. https://doi.org/10.3390/app11135925

Academic Editor: Sebastian Jaksch

Received: 1 June 2021
Accepted: 23 June 2021
Published: 25 June 2021

Publisher's Note: MDPI stays neutral with regard to jurisdictional claims in published maps and institutional affiliations.

Copyright: © 2021 by the authors. Licensee MDPI, Basel, Switzerland. This article is an open access article distributed under the terms and conditions of the Creative Commons Attribution (CC BY) license (https://creativecommons.org/licenses/by/4.0/).

1. Introduction

Small angle scattering (SAS) of X-rays (SAXS) and neutrons (SANS) are non-destructive powerful techniques which have been utilized successfully for investigating nanostructured materials such as metal nanoparticles [1,2], emulsions [3], micelles [4,5], electrolytes [6,7], liquid crystals [8,9] and organic nanoparticles [10–13]. The SAS technique provides nanoscale information including shape, size, and size distribution as well as spatial distribution of dispersed materials in solution [14,15]. In particular, they provide the diameter of gyration and the mass of isotropically distributed shaped nanoscale particles [16], as well as the linear mass density and cross-sectional size of anisotropically shaped particles [14,17]. Moreover, SAXS and SANS methods are important tools for examining materials for various applications, e.g., solar cells, lithium-ion batteries and sensors [6,7,18–21]. Combining both SAXS and SANS methods to investigate the same sample volume is remarkably advantageous, where two radiations offer two different contrast situations (X-ray and neutron scattering cross section). Using such multiple modalities is very beneficial, especially for cases when a single modality provides incomplete or ambiguous results. For instance, SANS provides the form factor of the shell in a representative hollow-shell-like particle [2]. In contrast, SAXS measurements of the same core–shell sample are sensitive to the high-electron-density core. The structures of both shell and core can be uniquely obtained from the same sample volume via complementary SAXS and SANS methods. Thus, simultaneous SAXS/SANS measurements will allow the exactness of the probed samples and the potential for using a common analysis model for both SAXS and SANS data, and consequently, a straightforward and unambiguous interpretation of the scattering data. For instance, structural modifications of supramolecular gels upon switching between H_2O and D_2O have been reported [22]. The deuteration effect on the structure of some nanoscale structured samples cannot be ignored. Thus, the simultaneous SAXS/SANS method is an essential tool for eliminating any ambiguity regarding sample exactness.

A key contribution of the simultaneous SAXS/SANS method is its ability to perform in situ and operando studies on advanced materials that undergo temporal modifications under external stimuli [23], allowing access to their structural rearrangements from the molecular scale to nanoscale assemblies. This means not only the synergy that results when data sets are obtained simultaneously from the same sample volume, but also access to the cross-correlation of phases/components when two different simultaneous contrast situations are probed for multi-phase/component samples. For instance, the cross-correlation between evolved gold nanorods size/shape (SAXS) and the corresponding engaged structure of organic moieties (SANS) during surfactant-assisted seed-mediated synthesis of gold nanorods will reveal new information related to the exact growth mechanisms involved [23]. Moreover, the SAXS/SANS system will enable contrast-dependent successive length-scale characterization of novel nanomaterials, addressing unsolved scientific questions.

A few instruments at neutron research reactors are currently utilized for both neutron and X-ray radiation experiment, obtaining complementary results of investigated samples. These dedicated instruments are mainly reflectometry and imaging techniques. A neutron imaging setup at PSI is outfitted with an additional X-ray tube to perform computerized tomography (CT) investigations using X-ray and neutron radiation [24,25]. Similarly, an instrument (NeXT) at ILL hosts two complementary imaging setups; neutron and X-ray tomography [26,27]. Neutron reflectometers such as the NREX instrument at MLZ allows combined X-ray and neutron specular reflectivity measurements using both X-ray and neutron beams placed in an orthogonal geometry [28]. Unfortunately, in most cases for the latter instruments, high energy background signal on the X-ray detector is a significant challenge and hampers a simultaneous mode of combined X-ray and neutron methods. The X-ray detection system is often not optimized for simultaneous acquisition, thus both X-ray and neutron data sets are mainly acquired successively and not in a synchronous mode [29,30].

An advanced SAXS system mounted on a standalone metal rack that makes it easily movable for use on the D22 SANS instrument has successfully been installed and tested at ILL [23]. For the first time, SAXS and SANS measurements in a simultaneous mode have been demonstrated [23]. One challenge is cutting down background signal that originates mainly from high-energy gamma radiation [31] in the experimental zone of the D22 instrument. A preliminary lead shielding design around our SAXS instrument at D22 of ILL has shown the capability of acquiring background discriminated SAXS data on a single energy threshold X-ray detector, enabling simultaneous SAXS/SANS experiments [23]. To enable a combined SAXS/SANS method in a simultaneous mode for all possible experimental conditions at the D22 instrument, an upgrade of the SAXS system was an essential requirement. In the current work, an optimized lead shielding has been implemented and the background suppression has been systematically examined. Additionally, a double energy threshold X-ray detector has been employed, replacing the single energy threshold detector. Overall reduced background noise and acceptable signal-to-noise ratios of SAXS data have been demonstrated upon combining both modifications, for all possible experimental configurations at the D22 instrument. Combined SAXS/SANS measurements were acquired for two nanoparticle (NP) dispersions of different sizes, to demonstrate the performance of the SAXS/SANS measurements under various experimental conditions.

2. Materials and Methods

An advanced SAXS system based on a Copper/Molybdenum (K_α; 8.0/17.4 keV) switchable microfocus rotating anode X-ray generator (Rigaku MM007 HF DW) and a Dectris EIGER2 X 1M detector with a continuously variable sample-to-detector distance ($SDD_{X\text{-ray}}$) from 0.55 m up to 1.63 m in a vacuum tube is dimensionally suitable for installation on the D22 SANS instrument at ILL (Figure 1). For Cu K_α radiation of 8 keV ($\lambda \approx 0.1540$ nm wavelength), the absolute value of the scattering vector q can be selected to range between 0.005 and 0.5 Å^{-1} by the $SDD_{X\text{-ray}}$. Due to the unsuitability of copper radiation for iron containing samples (such as magnetite, maghemite, and hematite) and for reaching higher penetration depths, the X-ray source can easily be switched to Mo $K\alpha$ radiation of 17.4 keV. The short wavelength (0.071 nm) of the Mo radiation also allows the q range of the SAXS instrument to be extended [23]. With a Rigaku VariMax™ beam divergence control (BDC) system mounted on the X-ray source, the dual wavelength (Cu and Mo) optics can be operated. A control software can be used to automate the VariMax DW™ optics when changing from one X-ray source to the other, so that the operation of the whole system can proceed easily. A collimation length of approximately 56 cm is composed of three different types of slits. The first tungsten carbide slit suitable for both Cu and Mo radiation is placed directly after the mirror on the source, however, two scatterless slits (JJ X-ray) are positioned at the end of the collimation. The last two scatterless slits, either silicon (used for Cu radiation) or GaAs (used for Mo radiation), define the divergence of the primary beam. The whole assembly (X-ray source, optics and collimation system) can be moved along three axes (x, y, z), enabling a fine tune of the X-ray beam position, so that both the X-ray and neutron beam can be superimposed at the same sample position. The beam size (FWHM) of 0.5 × 0.5 mm at the sample position provides a flux of approximately 1.5×10^7 photons/s (Cu source).

The components of the system are mounted on a standalone metal rack that makes it easily movable for use on the D22 instrument. The floor of the experimental zone is equipped with an adjustable metal support, allowing fast and precise positioning of SAXS system. Via a plug-and-play operation, the instrument can be used within half an hour following the installation process (1 h). NICOS control software (MLZ), combined with a TANGO environment, is used to control the whole instrument, including the motors and the SAXS data detection system. A goniometer (Huber) with six degrees of freedom is employed to achieve movements of the sample. The sample is typically positioned at 45° relative to both of the orthogonal X-ray and neutron beams. Following the sample position, an evacuated detector tube hosts a double energy threshold EIGER2 X 1M detector (pixel size = 75 × 75 μm^2). No beam stop is required to attenuate the direct beam, thus enabling investigations with direct beam intensity at the detector (no beam-stop shadow). We refer readers to our previous work [23] for more details on the instrument specifications.

Background intensity measurements at the X-ray detector were measured at different experimental conditions, including neutron collimation lengths and $SDD_{X\text{-ray}}$. Upon installing an optimized lead shielding arrangement on the metal chassis of SAXS setup, as well as employing a double energy threshold detector, the overall background reduction factor was evaluated compared with both the unshielded SAXS instrument and a single energy threshold detector. An extensive preliminary test prior to the latest instrumental upgrade was performed. The current results on background suppression demonstrate the achievements of increasing the signal-to-noise ratio on the X-ray detector. This was essential in order to enable a combined SAXS and SANS measurement in a synchronous mode. SANS data were collected using the D22 instrument at a neutron wavelength of 6 Å ($\Delta\lambda/\lambda = 10\%$), an aperture of 10 mm and $SDD_{Neutron}$ of 1.4, 5.6, 8, 17.6 m to cover a q-interval ranging from 3×10^{-3} to 0.7 Å^{-1}, corresponding to sizes $2R \sim \pi/q$ varying between 5 and 1050 Å. Aiming at examining the feasibility of acquiring an acceptably high signal-to-noise ratio for SAXS data at certain experimental conditions, two different sized silica NPs (100 nm and 26 nm) were tested, as these samples are of typical sizes (covering q-ranges) measured at various experimental conditions (collimation lengths 17.6,

8.0 and 5.6 m). The first sample was a powder of monodisperse silica NPs with a nominal size of 100 nm. The second sample was an aqueous dispersion of commercial silica NPs (Ludox TM50, 50 wt %) with an average diameter of approximately 26 nm, purchased from Sigma–Aldrich, Germany. A 5 mL solution (2 wt %) was prepared from each silica sample in a deuterated solvent (D_2O). Both SAXS and SANS data were simultaneously collected at different SDD_{X-ray}, $SDD_{neutron}$ and neutron collimation lengths. Both SAXS and SANS data sets were processed using Grasp software [32]. In the case of generating large numbers of SAXS and SANS data (for instance during in situ investigations), easy online data reduction and qualitative analysis of scattering data using this common software assists with the interactive optimization of experimental analysis.

3. Results and Discussion

A challenging requirement in order to perform experiments using combined X-ray and neutron radiation at large scale neutron research facilities is to shield the sensitive X-ray detector against the high-energy X-ray, neutron, and gamma radiation in the experimental zone. For high signal-to-noise ratio SAXS data, it is important to minimize the background radiation level, both from neutrons scattered by the sample and from gamma rays produced by neutron absorption within the sample. For this purpose, lead shielding and a double energy threshold detector were further employed.

3.1. Reduction of the Gamma Background by Lead Shielding

10 cm thick lead metal sheets were attached to the concrete walls towards the neighboring IN15 and WASP instruments to reduce the gamma radiation at the SAXS detector position in the D22 experimental zone. Additionally, several lead shielding walls were constructed and installed on the chassis of the SAXS instrument (Figure 1). The shielding walls include (i) a cone-shaped wall (nose) of lead-based material in front of the detector vacuum chamber, (ii) a lead wall along the final neutron collimation line, and (iii) a lead wall on the side of SAXS vacuum chamber. The current design has been pre-tested so that the X-ray detector inside the vacuum chamber is appropriately shielded from radiation for all possible positions of the X-ray detector in the detector tube. In the photograph of Figure 1, the installed lead-based shielding walls attached to the SAXS chassis are visible.

A double-energy threshold EIGER2 X 1M detector replacing a single-energy threshold EIGER R 1M detector is currently in operation and has been placed inside the vacuum chamber of the SAXS system. It allows the use of an upper and a lower adjustable energy discriminator (for instance, at 4 keV and 10 keV) to select a sensitive energy range close to the characteristic energy of the employed X-ray source (Cu K_α 8 keV).

Here, background measurements were recorded for two extreme configurations of neutron collimation lengths, referred to in the following as 'short' (2.8 m) and 'long' (17.6 m). The SAXS background measurements were performed inside the D22 zone using the single-energy (4 keV) and the double-energy threshold (4 and 10 keV) modes for both the lead shielded and the unshielded SAXS instrument. The unshielded system represents a configuration where the lead shielding walls i, ii, and iii (cf. Figure 1) are not installed. From the data visualized in Figure 2, it can be concluded that the detected background signal can be substantially reduced by the lead shielding for both neutron collimation lengths (2.8 and 17.6 m). Further reduction of the background can be achieved by switching the X-ray detector from the single-energy threshold mode (th1 mode: 4 keV, dashed area) to the difference mode (diff mode: 4–10 keV, solid area) (cf. Figure 2). Measurements were performed for both the short (SD1 = 0.55 m red bars) and the long (SD2 = 1.64 m blue bars) SSD_{X-ray} of the SAXS system. The achieved total background SFs are summarized in Table 1.

Figure 1. (**a**) Three lead shielding walls: (i) a cone of lead-based material, (ii) lead wall along the final part of the neutron collimation line, and (iii) a lead wall (2 tons) along the vacuum chamber of the X-ray detector. The shielding walls can be easily craned and then firmly attached to the metal chassis of the SAXS setup. Red and yellow arrows indicate the directions of the X-ray and neutron beams, respectively. The unshielded SAXS instrument (detector) represents a configuration where the lead shielding walls i, ii, and iii are not implemented during SAXS operation. A double-energy threshold X-ray detector was placed inside the vacuum tube, replacing the previously employed single-energy threshold detector. (**b**) A zoomed-in detail of the sample environment of the SAXS/SANS instrument, displaying a sample at an angle of 45° relative to both of the orthogonal neutron and X-ray beams.

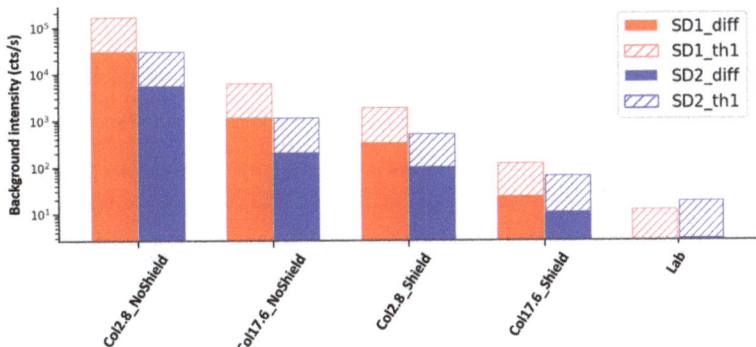

Figure 2. Background intensity collected using EIGER2 X 1M detector while neutron shutter is opened for two different neutron collimation lengths (2.8 m and 17.6 m). Two measurements for the single-energy threshold mode (dashed area, th1 mode = 4 keV) and two for the double-energy mode (solid area, diff = 4–10 keV) were performed for both the shielded and unshielded SAXS instrument. All background measurements were repeated for short (red, SD1 = 0.55 m) and long (blue, SD2 = 1.63 m) SDD_{X-ray}. Lab environment data are based on background intensity measurements while the reactor power was off. The intensity (cts/s) data are shown on a logarithmic scale.

Table 1. Background suppression factors (SFs) for all investigated experimental configurations in an optimized lead shielding setup with a double-energy threshold energy detector (total SF ≈ lead shielding SF × detector SF).

SDD$_{X\text{-ray}}$ (m)	Neutron Collimation Length (m)	X-ray Energy Mode (keV)	Shielding Suppression Factor	Detector Suppression Factor	Total Suppression Factor
0.55	2.8	4	87	-	87
		4–10	89	5.5	505
	17.6	4	58	-	58
		4–10	54	5.0	275
1.64	2.8	4	50	-	50
		4–10	49	5.1	248
	17.6	4	17	-	17
		4–10	19	6.1	116

The gamma background is originated from the facility environment and nuclear reactions of the neutrons with the sample material. The short collimation length of the SANS instrument (2.8 m) leads to a higher neutron flux compared to longer collimation lengths. The increase of neutron flux and the gamma radiation background level is roughly proportional to the inverse of the squared collimation length. To shield the X-ray detector (at different SDD$_{X\text{-ray}}$) against this high-energy gamma radiation, several preliminary tests were performed to choose the best lead wall dimensions and geometric arrangements for our SAXS system. The detector SF was calculated as the ratio of background intensity for single-energy and double-energy threshold modes. The total suppression factor is theoretically equal to the product of both shielding and detector SFs. An uncertainty of the total SF of ≤ 1% was observed under identical experimental conditions. Finally, large SFs were achieved, especially at short SDD$_{X\text{-ray}}$ (cf. Table 1). For SDD$_{X\text{-ray}}$ = 0.55 m, the background intensity could be reduced by a factor of 505 and 275 for short (2.8 m) and long (1.7.6 m) neutron collimation lengths, respectively. For both the unshielded and shielded instrument, the background SF for the double-energy threshold detection was almost equal. The gamma background which originates from the neutron interaction with the sample or from the environment and transmitted through the sample towards the detector cannot be shielded. But the double-energy threshold detector is able to reduce this background component. The tremendous reduction in the background signal of the upgraded instrument for the benefit of simultaneous SAXS/SANS measurements is demonstrated in Section 3.3 by SAXS/SANS data from NP samples.

3.2. Shielding the X-ray Source against External Magnetic Fields

It has occasionally been observed that the X-ray beam intensity periodically varies during SAXS measurements. Such beam intensity fluctuations had not been observed before operating it at the D22 SANS instrument. Using a slit size of 0.5 × 0.5 mm^2 in front of the sample, it was observed that the X-ray beam moves partially outside the slit's openings, leading to an intensity reduction of up to 60%. It was a challenge to identify a reason behind such occasional instability behavior of X-ray intensity during operation inside the experimental zone. Finally, the variation of the residual weak magnetic field (~2–5 Gauss) from the IN15 spin-echo spectrometer was found to cause these X-ray beam intensity fluctuations. A plate of soft ferromagnetic alloy was installed to shield the static and low-frequency magnetic field at the position of the X-ray source, which leads to a stable beam intensity (Figure 3) for all extreme magnetic fields employed at the IN15 instrument.

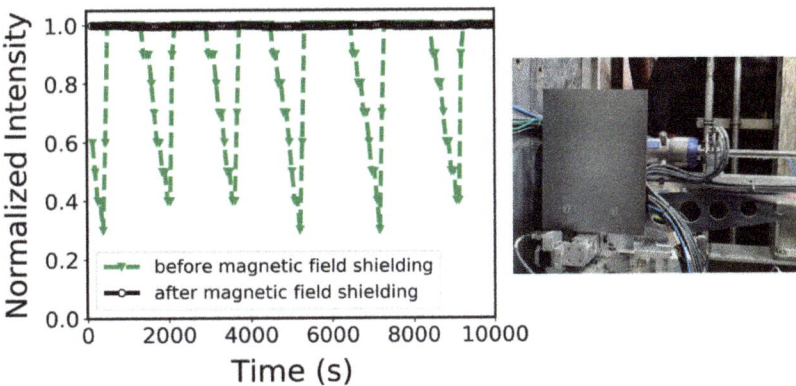

Figure 3. X-ray beam intensity before (no magnetic shielding, green line) and after (magnetic field shielding, black symbols) installing a 1 mm thick plate of a soft ferromagnetic alloy at the X-ray source. A magnetic field strength of approximately 2–5 Gauss at the X-ray source causes an X-ray beam drift.

3.3. Overall SAXS Performance

For simultaneous SAXS/SANS experiments, it is important to achieve sufficient data quality in a given measuring time for both methods. For that purpose, it was important to reduce the background signal originating from gamma rays on the 2D X-ray detector of the SAXS instrument for all experimental conditions represented by different SANS collimation lengths and primary neutron beam fluxes and sizes, as well as different SDD_{X-ray}. As demonstrated above, this could be achieved by a targeted enhancement of the neutron and magnetic field shielding, as well as using a double-energy threshold X-ray detector. To demonstrate the success of these upgrades for the performance of the SAXS instrument, two NP samples of nominal sizes of 100 nm and 26 nm were studied by simultaneous SAXS/SANS experiments. The scattering data was collected while the nuclear reactor was in operation (neutron shutter opened, reactor power = 43 MW). SAXS data of the same samples were also collected when the neutron reactor was in the shutdown time (neutron shutter closed, reactor power < 1.5 MW). The latter experiments allow a comparison of the SAXS data of the investigated samples under simultaneous SAXS/SANS data recording conditions with standard laboratory standalone SAXS experiments, where only natural cosmic background contributes to the gamma background of the SAXS data.

SAXS data of silica NP samples were collected for 15 min for both short (0.55 m) and long (1.63 m) SDD_{X-ray} (Figure 4). For long SDD_{X-ray}, SANS measurements of both silica NPs were simultaneously acquired at a long collimation length of 17.6 m. For short SDD_{X-ray}, SANS data were simultaneously measured at two different collimation lengths of 8 m and 5.6 m for 100 nm and 26 nm silica NPs, respectively. The SAXS 2D detector patterns were corrected and normalized, respectively, for measuring time, sample thickness, transmission, flat field, solid angle covered by each detector pixel, and incident intensity. The data were calibrated to absolute intensities, and the q-scale was azimuthally averaged. Background (dark current) correction was performed by subtracting the background intensity measured on the detector at relevant operating conditions. For each neutron collimation length used, the background intensity on the X-ray detector was collected for several minutes while the neutron shutter was opened but the X-ray shutter was closed. Besides this, each SAXS data set was corrected for a constant background determined from a measurement of the corresponding sample with a closed neutron shutter (and reactor power < 1.5 MW).

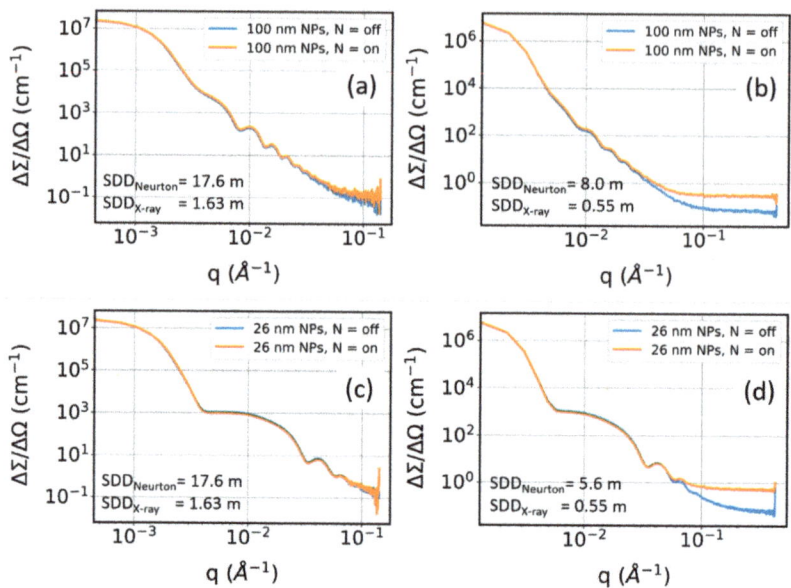

Figure 4. 1D SAXS profiles at various neutron collimation lengths and $SDD_{X\text{-ray}}$ of (**a**,**b**) 100 nm and (**c**,**d**) 26 nm silica NPs collected with the neutron beam switched on (orange curves) or off (blue) (simultaneous and asynchronous modes). The asynchronous mode represents a standard laboratory SAXS experimental condition where neutron shutter was closed and the nuclear reactor was almost switched off (below < 1.5 MW). SAXS/SANS simultaneous mode represents an experimental condition where the neutron shutter of D22 was open (reactor power = 43 MW). At long $SDD_{X\text{-ray}}$, the SAXS profiles in simultaneous mode are well matching those of the asynchronous mode, which leads to a good signal-to-noise ratio for the SAXS data. For short collimations, noticeable background which was approximately 4 and 8 times larger than that for standard lab SAXS measurements was found for both collimations 8 m (**b**) and 5.6 m (**d**), respectively.

At a long collimation length (17.6 m), the SAXS 1D profiles are well matching those collected in absence of the high-energy gamma radiation environment (Figure 4a,c). This reveals that the SAXS data quality of a simultaneous SAXS/SANS measurement at those conditions (collimation length = 17.6 m, $SDD_{X\text{-ray}}$ = 1.63 m) is not compromised with respect to the background level of a state-of-the-art lab-based SAXS measurement. At a short distance $SDD_{X\text{-ray}}$ (0.55 m), SANS images were collected in a synchronous mode at collimation lengths of 8 m (for 100 nm NPs) and 5.6 m (for 26 nm NPs). In the latter configurations, the SAXS profiles exhibited a background intensity approximately 4 and 8 times higher than the standard lab-based SAXS measurements for both 8 m and 5.6 m collimation lengths, respectively (Figure 4b,d). This gamma background originated predominantly from the gamma rays produced by neutron capture within the sample. Nevertheless, the overall background level even for the short neutron collimation lengths is still low and allows high quality simultaneous SAXS/SANS experiments. Based on the q-range of characteristic interference features, scattering power of investigated samples, and experimental conditions (such as collimation length, $SDD_{X\text{-ray}}$, X-ray slit size, and measuring time), one needs to carefully optimize the experimental conditions to achieve the best quality simultaneous SAXS/SANS experiments.

Luckily, the D22 instrument is currently equipped with a second fixed detector at 1.4 m from the sample position in addition to a movable detector, enabling extended SANS q ranges (from 0.003 up to 0.7 Å^{-1} in a single configuration). Hence, typical SANS experiments are performed with collimation lengths > 5 m. Thus, the high signal-to-background level of the SAXS data detected at different collimation lengths (17.6, 8 and 5.6 m) allows state-of-the-art time-resolved in situ simultaneous SAXS/SANS measurements. This has been demonstrated by studies on the growth of gold nanorods using surfactant-assisted seed-mediated synthesis using simultaneous SAXS/SANS at a neutron collimation length of 8 m [23].

For SAXS/SANS data evaluation of the studied NP dispersions (Figure 5), a simple sphere model with a log-normally distributed radius coupled with a distribution width parameter implemented in SASfit software [33] was used. In SASfit, simultaneous fits of the model to the SAXS and SANS experimental data sets were performed with a convergence $\chi \ll 1$. The scattering length density contrast $\Delta SLD = SLD_{NPs} - SLD_{solvent}$ was fixed to the known values for silica to be 13.3 and 2.2 Å^{-2} for SAXS and SANS, respectively. The data can be nicely fitted with the simple model and reveals a particle diameter (with a distribution width) of 110 $\pm$ 7 nm (nominal 100 nm) and 26 $\pm$ 3 nm (nominal 26 nm), respectively. This simple example demonstrates that a simultaneous fit of a structural model to the SAXS and SANS data set can be easily performed on an absolute intensity scale and reveal the structural parameters, circumventing ambiguities which might be left if only one contrast (X-ray or neutron) is used. This is a particular advantage for in situ or operando studies of complex nanoscale systems which slowly change their structure, e.g., upon a reaction, crystallization, aggregation, gelation, particle formation, growth or aging process.

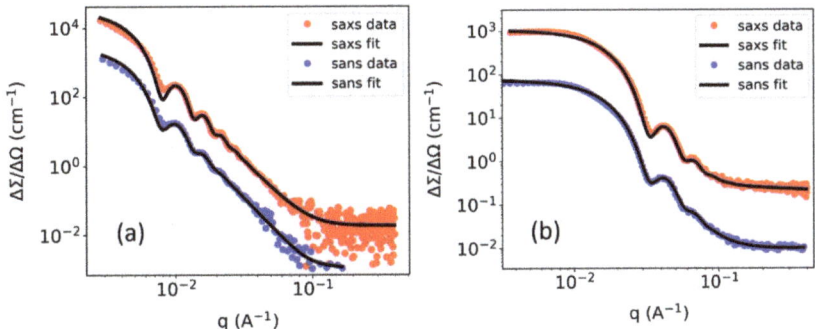

Figure 5. Solvent subtracted SAXS/SANS intensity profiles of 100 nm (**a**) and 26 nm (**b**) silica NPs acquired in a simultaneous mode. A common structural model is used for each sample to simultaneously fit both SAXS and SANS data collected from the same sample volume.

4. Conclusions

A customized SAXS system was successfully installed and operated in combination with the D22 SANS instrument of ILL, resulting in simultaneously collected SAXS and SANS data of the same part of a sample. Due to high-energy gamma radiation inside the D22 instrumental zone, the SAXS instrument was upgraded with a double-energy threshold detector and optimized lead shielding. Background suppression factors were evaluated for different collimation lengths and SDD$_{X-ray}$. Extreme background reduction by factors up to 505 was achieved for short collimation lengths and SDD$_{X-ray}$. Acceptable signal-to-background ratios of the SAXS patterns comparable to SAXS data collected in the absence of gamma background radiation could be achieved. For long neutron collimation lengths of up to 17.6 m and long SDD$_{X-ray}$, the SAXS profiles did not exhibit an increase in background at all when compared to a typical lab experiment. An increase of the corresponding low background by factors of 4 and 8 were observed for neutron collimation lengths of 8 m and 5.6 m, respectively. In the latter experimental configurations, SAXS data of selected test samples prove the suitability of the SAXS instrument for combined SAXS/SANS experiments. The upgraded SAXS system is now compatible with all collimation settings of the D22 SANS instrument, enabling in situ and operando simultaneous SAXS/SANS measurements for a variety of complex nanoscale sample systems.

Author Contributions: Conceptualization, T.U., C.B., E.M., K.G., T.Z., L.P.; methodology, E.M., K.G., T.Z., L.P., I.S., C.B., A.M.; data curation, E.M., I.S.; formal analysis, E.M., I.S.; writing—original draft preparation, E.M.; writing—review and editing T.U., L.P., A.M., I.S.; technical details, C.B., E.M., T.U., L.P.; funding acquisition, T.U. All authors have read and agreed to the published version of the manuscript.

Funding: This research is primarily funded by the Federal Ministry of Education and Research of Germany (BMBF) in the framework of project No. 05K16WE1. As well, this research received funding from the Deutsche Forschungsgemeinschaft (DFG) through the 'Cluster of Excellence Engineering of Advanced Materials (EAM)', the research training group GRK 1896 'In-situ Microscopy with Electrons, X-rays and Scanning Probes' and the research unit FOR 1878 'Functional Molecular Structures on Complex Oxide Surfaces'.

Institutional Review Board Statement: Not applicable.

Informed Consent Statement: Not applicable.

Data Availability Statement: Data presented in this study can be obtained from the authors upon request.

Acknowledgments: We gratefully thank Herbert Lang (ICSP), Jürgen Grasser (ISCP), Mark Jacques (ILL) for their technical support. We acknowledge ILL for the beamtime DOI:10.5291/ILL-DATA.LTP-9-9.

Conflicts of Interest: The authors declare no conflict of interest.

Abbreviations

The following abbreviations are used in this manuscript:

SDD_{X-ray}	Sample-to-detector distance of SAXS
$SDD_{Neutron}$	Sample-to-detector distance of SANS
SLD	Scattering length density
col	Neutron collimation
diff	Difference mode of double-energy threshold X-ray detector, 4–10 KeV
th1	A single-energy threshold mode of X-ray detector, 4 KeV
SD1	Sample-to-detector distance of SAXS at 0.55 m
SD2	Sample-to-detector distance of SAXS at 1.632 m
shield	Lead shielding on the SAXS system
NoShield	SAXS system without any lead shielding walls

References

1. Schmutzler, T.; Schindler, T.; Zech, T.; Lages, S.; Thoma, M.; Appavou, M.S.; Peukert, W.; Spiecker, E.; Unruh, T. n-Hexanol Enhances the Cetyltrimethylammonium Bromide Stabilization of Small Gold Nanoparticles and Promotes the Growth of Gold Nanorods. *ACS Appl. Nano Mater.* **2019**, *2*, 3206–3219. [CrossRef]
2. Schindler, T.; Lin, W.; Schmutzler, T.; Lindner, P.; Peukert, W.; Segets, D.; Unruh, T. Evolution of the Ligand Shell Around Small ZnO Nanoparticles During the Exchange of Acetate by Catechol: A Small Angle Scattering Study. *ChemNanoMat* **2019**, *5*, 116–123. [CrossRef]
3. Schmiele, M.; Busch, S.; Morhenn, H.; Schindler, T.; Schmutzler, T.; Schweins, R.; Lindner, P.; Boesecke, P.; Westermann, M.; Steiniger, F.; et al. Structural Characterization of Lecithin-Stabilized Tetracosane Lipid Nanoparticles. Part I: Emulsions. *J. Phys. Chem. B* **2016**, *120*, 5505–5512. [CrossRef]
4. Schmutzler, T.; Schindler, T.; Goetz, K.; Appavou, M.S.; Lindner, P.; Prevost, S.; Unruh, T. Concentration dependent morphology and composition of n-alcohol modified cetyltrimethylammonium bromide micelles. *J. Phys. Condens. Mat.* **2018**, *30*. [CrossRef] [PubMed]
5. Schmutzler, T.; Schindler, T.; Schmiele, M.; Appavou, M.S.; Lages, S.; Kriele, A.; Gilles, R.; Unruh, T. The influence of n-hexanol on the morphology and composition of CTAB micelles. *Colloid Surface A* **2018**, *543*, 56–63. [CrossRef]
6. Mohl, G.E.; Metwalli, E.; Bouchet, R.; Phan, T.N.T.; Cubitt, R.; Muller-Buschbaum, P. In Operando Small-Angle Neutron Scattering Study of Single-Ion Copolymer Electrolyte for Li-Metal Batteries. *ACS Energy Lett.* **2018**, *3*, 1–6. [CrossRef]
7. Mohl, G.E.; Metwalli, E.; Muller-Buschbaum, P. In Operando Small-Angle X-ray Scattering Investigation of Nanostructured Polymer Electrolyte for Lithium-Ion Batteries. *ACS Energy Lett.* **2018**, *3*, 1525–1530. [CrossRef]

8. Gehrer, S.; Schmiele, M.; Westermann, M.; Steiniger, F.; Unruh, T. Liquid Crystalline Phase Formation in Suspensions of Solid Trimyristin Nanoparticles. *J. Phys. Chem. B* **2014**, *118*, 11387–11396. [CrossRef] [PubMed]
9. Schmiele, M.; Gehrer, S.; Unruh, T. Small-angle scattering simulations for suspensions of nanocrystals. *Acta Crystallogr. A* **2014**, *70*, C597. [CrossRef]
10. Schuldes, I.; Noll, D.M.; Schindler, T.; Zech, T.; Gotz, K.; Appavou, M.S.; Boesecke, P.; Steiniger, F.; Schulz, P.S.; Unruh, T. Internal Structure of Nanometer-Sized Droplets Prepared by Antisolvent Precipitation. *Langmuir* **2019**, *35*, 13578–13587. [CrossRef]
11. Unruh, T. Interpretation of small-angle X-ray scattering patterns of crystalline triglyceride nanoparticles in dispersion. *J. Appl. Crystallogr.* **2007**, *40*, 1008–1018. [CrossRef]
12. Schmiele, M.; Schindler, T.; Westermann, M.; Steiniger, F.; Radulescu, A.; Kriele, A.; Gilles, R.; Unruh, T. Mesoscopic Structures of Triglyceride Nanosuspensions Studied by Small-Angle X-ray and Neutron Scattering and Computer Simulations. *J. Phys. Chem. B* **2014**, *118*, 8808–8818. [CrossRef] [PubMed]
13. Wibmer, L.; Lages, S.; Unruh, T.; Guldi, D.M. Excitons and Trions in One-Photon- and Two-Photon-Excited MoS2: A Study in Dispersions. *Adv. Mater.* **2018**, *30*. [CrossRef] [PubMed]
14. Putnam, C.D.; Hammel, M.; Hura, G.L.; Tainer, J.A. X-ray solution scattering (SAXS) combined with crystallography and computation: Defining accurate macromolecular structures, conformations and assemblies in solution. *Q. Rev. Biophys.* **2007**, *40*, 191–285. [CrossRef] [PubMed]
15. Glatter, O.; Kratky, O. *Small Angle X-ray Scattering*; Academic Press: London, UK; New York, NY, USA, 1982; p. 515.
16. Serdyuk, I.N.; Tsalkova, T.N.; Svergun, D.I.; Izotova, T.D. Determination of Radii of Gyration of Particles by Small-Angle Neutron-Scattering—Calculation of the Effect of Aggregates—Appendix. *J. Mol. Biol.* **1987**, *194*, 126–128. [CrossRef]
17. Koch, M.H.J.; Vachette, P.; Svergun, D.I. Small-angle scattering: A view on the properties, structures and structural changes of biological macromolecules in solution. *Q. Rev. Biophys.* **2003**, *36*, 147–227. [CrossRef]
18. Schindler, T.; Walter, J.; Peukert, W.; Segets, D.; Unruh, T. In Situ Study on the Evolution of Multimodal Particle Size Distributions of ZnO Quantum Dots: Some General Rules for the Occurrence of Multimodalities. *J. Phys. Chem. B* **2015**, *119*, 15370–15380. [CrossRef]
19. Futscher, M.H.; Schultz, T.; Frisch, J.; Ralaiarisoa, M.; Metwalli, E.; Nardi, M.V.; Muller-Buschbaum, P.; Koch, N. Electronic properties of hybrid organic/inorganic semiconductor pn-junctions. *J. Phys. Condens. Mat.* **2019**, *31*. [CrossRef]
20. Wang, X.Y.; Meng, J.Q.; Wang, M.M.; Xiao, Y.; Liu, R.; Xia, Y.G.; Yao, Y.; Metwalli, E.; Zhang, Q.; Qiu, B.; et al. Facile Scalable Synthesis of TiO2/Carbon Nanohybrids with Ultrasmall TiO2 Nanoparticles Homogeneously Embedded in Carbon Matrix. *ACS Appl. Mater. Inter.* **2015**, *7*, 24247–24255. [CrossRef]
21. Wang, X.Y.; Zhao, D.; Wang, C.; Xia, Y.G.; Jiang, W.S.; Xia, S.L.; Yin, S.S.; Zuo, X.X.; Metwalli, E.; Xiao, Y.; et al. Role of Nickel Nanoparticles in High-Performance TiO2/Ni/Carbon Nanohybrid Lithium/Sodium-Ion Battery Anodes. *Chem. Asian J.* **2019**, *14*, 2169. [CrossRef]
22. McAulay, K.; Wang, H.; Fuentes-Caparros, A.M.; Thomson, L.; Khunti, N.; Cowieson, N.; Cui, H.G.; Seddon, A.; Adams, D.J. Isotopic Control over Self-Assembly in Supramolecular Gels. *Langmuir* **2020**, *36*, 8626–8631. [CrossRef]
23. Metwalli, E.; Gotz, K.; Lages, S.; Bar, C.; Zech, T.; Noll, D.M.; Schuldes, I.; Schindler, T.; Prihoda, A.; Lang, H.; et al. A novel experimental approach for nanostructure analysis: Simultaneous small-angle X-ray and neutron scattering. *J. Appl. Crystallogr.* **2020**, *53*, 722–733. [CrossRef]
24. Deschler-Erb, E.; Lehmann, E.H.; Pernet, L.; Vontobel, P.; Hartmann, S. The complementary use of neutrons and X-rays for the non-destructive investigation of archaeological objects from Swiss collections. *Archaeometry* **2004**, *46*, 647–661. [CrossRef]
25. Lehmann, E.H.; Mannes, D.; Kaestner, A.P.; Hovind, J.; Trtik, P.; Strobl, M. The XTRA Option at the NEUTRA Facility—More Than 10 Years of Bi-Modal Neutron and X-ray Imaging at PSI. *Appl. Sci.* **2021**, *11*, 3825. [CrossRef]
26. Robuschi, S.; Tengattini, A.; Dijkstra, J.; Fernández, I.; Lundgren, K. A closer look at corrosion of steel reinforcement bars in concrete using 3D neutron and X-ray computed tomography. *Cem. Concr. Res.* **2021**, *144*, 106439. [CrossRef]
27. Ziesche, R.F.; Arlt, T.; Finegan, D.P.; Heenan, T.M.M.; Tengattini, A.; Baum, D.; Kardjilov, N.; Markotter, H.; Manke, I.; Kockelmann, W.; et al. 4D imaging of lithium-batteries using correlative neutron and X-ray tomography with a virtual unrolling technique. *Nat. Commun.* **2020**, *11*. [CrossRef] [PubMed]
28. Khaydukov, Y.; Soltwedel, O.; Keller, T. NREX: Neutron reflectometer with X-ray option. *J. Large-Scale Res. Facil.* **2015**, *1*, A38. [CrossRef]
29. Mannes, D.; Schmid, F.; Frey, J.; Schmidt-Ott, K.; Lehmann, E. Combined Neutron and X-ray imaging for non-invasive investigations of cultural heritage objects. *Phys. Procedia* **2015**, *69*, 653–660. [CrossRef]
30. Makarova, M.V.; Kravtsov, E.A.; Proglyado, V.V.; Khaydukov, Y.; Ustinov, V.V. Structure and Magnetism of Co/Dy Superlattices. *Phys. Solid State* **2020**, *62*, 1664–1666. [CrossRef]
31. Murray, E.; Smith, A.G.; Pollitt, A.J.; Matarranz, J.; Tsekhanovich, I.; Soldner, T.; Koster, U.; Biswas, D.C. Measurement of Gamma Energy Distributions and Multiplicities Using STEFF. *Nucl. Data Sheets* **2014**, *119*, 217–220. [CrossRef]
32. Dewhurst, C. GRASP. Available online: https://www.ill.eu/users/support-labs-infrastructure/software-scientific-tools/grasp (accessed on 15 March 2021).
33. Bressler, I.; Kohlbrecher, J.; Thunemann, A.F. SASfit: A tool for small-angle scattering data analysis using a library of analytical expressions. *J. Appl. Crystallogr.* **2015**, *48*, 1587–1598. [CrossRef] [PubMed]

Communication

Feasibility of Probing the Filler Restructuring in Magnetoactive Elastomers by Ultra-Small-Angle Neutron Scattering

Inna A. Belyaeva [1], Jürgen Klepp [2,*], Hartmut Lemmel [3,4] and Mikhail Shamonin [1]

[1] East Bavarian Centre for Intelligent Materials (EBACIM), Ostbayerische Technische Hochschule (OTH) Regensburg, Prüfeninger Str. 58, 93049 Regensburg, Germany; inna.belyaeva@oth-regensburg.de (I.A.B.); mikhail.chamonine@oth-regensburg.de (M.S.)
[2] Faculty of Physics, University of Vienna, Boltzmanngasse 5, 1090 Vienna, Austria
[3] Atominstitut, Technische Universität Wien, 1020 Vienna, Austria; lemmel@ill.fr
[4] Institut Laue-Langevin, 38042 Grenoble, France
* Correspondence: juergen.klepp@univie.ac.at

Abstract: Ultra-small-angle neutron scattering (USANS) experiments are reported on isotropic magnetoactive elastomer (MAE) samples with different concentrations of micrometer-sized iron particles in the presence of an in-plane magnetic field up to 350 mT. The effect of the magnetic field on the scattering curves is observed in the scattering vector range between 2.5×10^{-5} and 1.85×10^{-4} Å^{-1}. It is found that the neutron scattering depends on the magnetization history (hysteresis). The relation of the observed changes to the magnetic-field-induced restructuring of the filler particles is discussed. The perspectives of employing USANS for investigations of the internal microstructure and its changes in magnetic field are considered.

Keywords: ultra-small-angle neutron scattering; magnetoactive elastomer; magnetorheological elastomer; hysteresis; restructuring of the filler

1. Introduction

Magnetoactive elastomers (MAEs) can be classified as magnetic composite materials, i.e., polymer matrices filled with magnetic micro-particles. The growing interest in MAEs is determined by the possibility of controlling their mechanical and other physical properties by application of technically easy realizable magnetic fields [1–9]. The physical reason for the observed changes in MAE properties are field-induced reconfigurations of the microstructure formed by the magnetic particles [1,5,10–12], which are possible in mechanically compliant polymer matrices [8]. These rearrangements of filling particles in an externally applied magnetic field have been observed directly by optical [13] and X-Ray methods [14,15].

The internal filler structure of nanocomposites with magnetic particles has been studied by small-angle neutron scattering (SANS) in the past [16]. Hitherto, most of SANS research has concentrated on ferrofluids [16–19] which are highly dispersed systems of nanometer-sized ferromagnetic or ferrimagnetic particles in a liquid medium (organic or inorganic fluids, water). Recently, first papers appeared, where magnetizable nano- [20–23] and micro-particles [21] embedded into an elastomer matrix have been studied by SANS in the scattering vector Q range ~0.01–0.1 Å^{-1}. Such scattering vectors do not correspond to the size of filling micro-particles. Balasiou et al. [21] have found that the SANS scattering intensities of an elastomer matrix with iron (Fe) micro-particles exhibited power law behaviour $I(Q) \approx Q^{-\alpha}$ with the exponent $3 < \alpha < 4$. For elastomer samples with Fe particle concentration between 25 and 75 mass%, a surface fractal dimension of $D_s \approx 2.2$ was obtained. The measurements were performed in the absence of a magnetic field. A theoretical approach to extract information about the internal structure of magnetic filament brushes from scattering experiments has been recently proposed by Pyanzina et al. [24].

Borin et al. [25] developed a Small Angle Light Scattering (SALS) laboratory setup with simultaneous exposure of fluidic samples to a shear flow and an external magnetic field and used it for investigation of structuring processes in a magnetic fluid comprising clustered iron oxide nanoparticles. They observed an anistropy of the scattering patterns in an applied magnetic field. Zákutná et al. [26] reported a magnetorheological SANS setup at the Institut Laue-Langevin allowing for an in-situ magnetic field up to 20 mT and the shear flow of ferrofluids. The maximum magnetic field in this setup is one order of magnitude below the magnetic-field values required to observe the effects caused by magnetic-field restructuring in MAEs.

Conventional SANS is an experimental technique that uses elastic neutron scattering at small scattering angles to investigate the structure of various substances at a mesoscopic scale of about 1–100 nm. Therefore, it has been neglected for investigation of MAEs, where micro-particles are employed. In the present paper, we use the ultra-small-angle neutron scattering (USANS) instrument S18 at the Institut Laue-Langevin where the length scale that can be analyzed is extended beyond 10 μm.

The displacements of magnetic particles in MAEs can be observed by using X-ray micro-computed topography, but this method becomes unreliable at high concentrations of filler particles due to X-ray absorption. It is known that SANS allows one to determine the fractal organization of a cluster (mass fractal) because a certain correlation between particles occurs in the case of their aggregation. It is also largely accepted that the reduction of the shear storage modulus of filled elastomers with the growing amplitude of shearing oscillations (known as the Payne effect) in nanocomposites can be attributed to the agglomeration/de-agglomeration of filler-particle aggregates and is determined by the fractal dimension of the cluster. Very recently, it was hypothesized that USANS may be used to determine the magnetic-field-induced changes of fractal dimension of aggregates of ferromagnetic particles in MAEs [27]. In general, it can be expected that if a rearrangement of magnetized inclusions takes place in a magnetic field, such a change in a microstructure should be observable in a USANS experiment if the size of particles or their aggregates is comparable with the magnitude of the observable scattering vector. Given some similarity between operation modes of MAEs and magnetic fluids, it would be interesting to obtain additional information on the internal microstructure of MAEs using neutron scattering techniques, as it has been successfully done in magnetic fluids. This Communication reports the first USANS observation in MAEs in the scattering vector range 2.5×10^{-5}–1.85×10^{-4} Å^{-1} (corresponding to structure sizes between 25 μm and 3.4 μm) and the effect of magnetic fields on the scattering curves.

2. Materials and Methods
2.1. Samples

Six thin-film MAE samples have been synthesized. The fabrication of MAE samples is described in detail in [27,28]. All samples have been cross-linked in the absence of a magnetic field and, therefore, it is known that they can be considered isotropic. The soft-magnetic carbonyl iron powder (CIP) type SQ (mean particle size d50 of 4.5 μm, no coating), provided by BASF SE Carbonyl Iron Powder & Metal Systems (Ludwigshafen, Germany), was used as the ferromagnetic filler.

Table 1 presents the iron content in three samples that had a soft elastomer as a matrix (the shear storage modulus is ~1 kPa). The thickness of MAE samples was 0.3 ± 0.1 mm. The thickness of the matrix (empty) sample was 0.7 mm. Measurements of SANS scattering curves of the matrix in the absence as well as in the presence of a magnetic field did not reveal any significant effect of magnetic field on the course of curves. We also investigated three MAE samples with 60 and 67 mass% of Fe which have been fabricated with a stiffer polymer matrix (the shear storage modulus is ~10 kPa). There one may expect that the magnetic-field restructuring will be diminished by a stiffer matrix.

Table 1. Samples.

Sample	Mass Fraction of Fe, %	Volume Fraction of Fe, %
Matrix	0	0
MAE10	10	1.3
MAE30	30	4.9
MAE80	80	32.4

2.2. Measurements

The USANS data were measured at the S18 interferometry and USANS facility of the Institut Laue-Langevin, Grenoble, France (Figure 1).

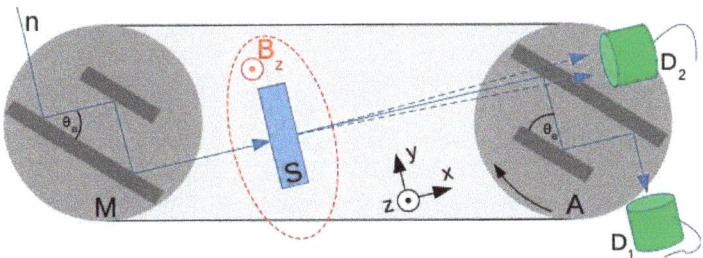

Figure 1. USANS setup (not to scale): so-called Bonse-Hart camera configuration with channel-cut silicon perfect crystals (monochromator M and analyzer A, lattice planes are parallel to the crystal slab faces), sample S in a region with an applied vertical, static magnetic field B_z, and detectors D_1 and D_2 [29].

The non-polarized neutron beam from the top left is diffracted by M to pass the sample S at normal incidence. Portions of the beam that are scattered by S will be diffracted by analyzer A and detected by detector D_1 at angular positions of A slightly different from the Bragg position θ_B. The accuracy of the piezo system controlling A's angular position is 0.036″. The Q-values in our USANS experiments refer to variation of A's angular position and are calculated from measured absolute piezo encoder values. Low-background conditions and the excellent suppression of diffraction-peak side lobes enable us to perform measurements at Q-values as small as a few 10^{-5} 1/Å. The wavelength of the incident beam was 1.9 Å. Neutrons are counted in detectors D_1 and D_2 as a function of angular position of A with and without sample. The values are normalized by the orders of magnitude larger count rate at a monitor detector upstream to account for power fluctuations of the neutron source (reactor) during the measurement. The scattering curve measured without sample was interpolated and subtracted from the curves measured with samples.

The dc magnetic field is applied vertically (z-direction) and it is in the plane of MAE samples. It can be varied from 0 to 350 mT. The neutron beam is directed perpendicular to the sample's plane and to the magnetic field vector. No visible deformation of MAE samples in the maximum magnetic field has been observed. The magnitude of the scattering vector Q refers to the horizontal plane. It can be expected that the restructuring (formation of elongated aggregates) occurs in the vertical direction and we mainly observe the scattering in the perpendicular direction. Unfortunately, due to the specific features of the USANS setup, it was not possible to record the 2D scattering patterns, because these measurements would be prohibitively long.

3. Results and Discussion

Although the data have been obtained in the Q-range up to 10^{-3} Å^{-1}, the following Figures show the data for the Q values up of 1.85×10^{-4} Å^{-1}. Even if we also suspect some magnetic-field-induced change in scattering intensity by structures smaller than

3.4 µm ($Q > 1.85 \times 10^{-4}$ Å^{-1}), we did not observe systematic and conclusive changes for larger Q values. Figure 2 presents the USANS curves in zero, intermediate (200 mT) and maximum (350 mT) magnetic field. There and in the following Figures the scattering curves are always normalized to the maximum of the presented curves (i.e., for a specific MAE sample). By such a normalization, the effect of magnetic field on neutron scattering in each particular MAE sample is demonstrated. We are currently not able to interpret the cross-correlation between scattering intensities of different samples, although such an information is available in the data set (cf. the data link at the end of this paper). The difference of the scattering curves is clearly seen and this difference grows with the increasing magnetic field. The total scattering intensity in the observed range of scattering vector is reduced with increasing magnetic flux density.

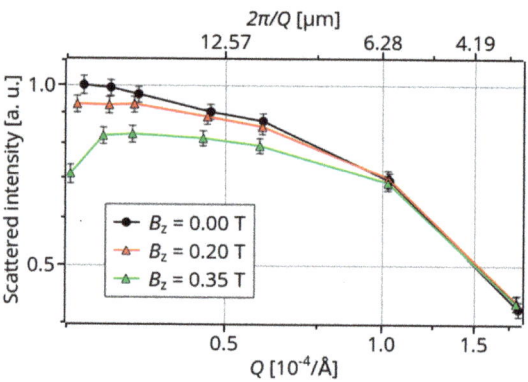

Figure 2. USANS curves of the MAE samples with 10 mass% of Fe for three different values of the magnetic flux density B_z.

Figure 3 shows the USANS curves for the sample with 30 mass% of Fe at different magnetic fields. Again, the effect of magnetic field is clearly observed. The scattering intensity becomes lower with the increasing magnetic field for 2.5×10^{-5} Å^{-1} < Q < 1×10^{-4} Å^{-1}.

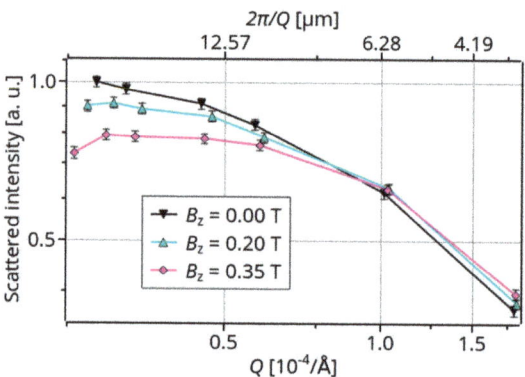

Figure 3. USANS curves of the MAE samples with 30 mass% of Fe for different magnetic fields.

As a control experiment, we measured the magnetic field dependence of the scattering curve of available MAE samples with the stiff matrix.

Figure 4 shows as an example the results for the MAE sample with 67 mass% of iron. A rather constant minor offset of the curve in the largest magnetic field appears to be within the measurement accuracy, so that no significant effect of magnetic field on the neutron scattering in the Q-range of interest is observed. In the current configuration both the nuclear and magnetic scattering cross sections are probed as Q is perpendicular to the magnetization direction. However, if magnetic scattering was present in this case, we were not able to reliably separate its contribution applying rather low magnetic field and using the experimental configuration shown in Figure 1.

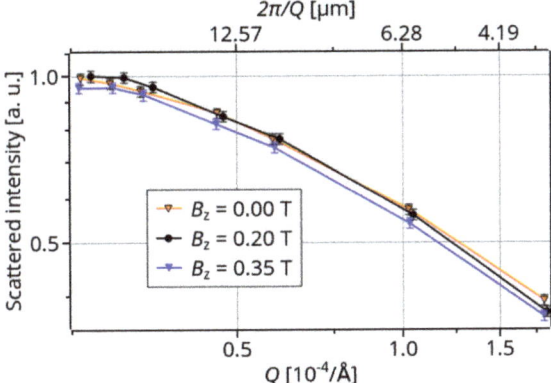

Figure 4. USANS curves of the MAE sample with 67 mass% of Fe and a stiff matrix for different magnetic fields.

Such an observation could be attributed to the limited mobility of filler micro-particles within the stiff matrix (as compared to the other MAE samples). The restructuring of filler particles is hindered in this case and the scattering behaviour is much less changed in comparison to other samples.

It is well known that physical properties of compliant MAE demonstrate hysteresis behaviour with respect to an applied magnetic field. The hysteresis occurs not only when the matrix is filled with hard-magnetic particles (where a kind of hysteresis has to be expected) but also for soft-magnetic inclusions with the vanishing magnetization reversal loop, which is the case in the present paper. Zubarev et al. [30] explained the hysteresis of magnetic properties by the hysteresis of the consolidation of filler particles into elongated aggregates. Since it is expected that the hysteresis is determined by different microstructures in MAEs depending on the magnetization history, it can be expected that the hysteresis can be observed in the scattering curves.

Figure 5 demonstrates the results of the scattering measurements. The following sequence of magnetic field strength was applied 0 T (denoted as (1) in Figure 5) → 0.2 T (2) → 0.35 T (3) → 0.2 T (4) → 0 T (5) → 0.2 T (6) → 0.35 T (7) → 0.2 T (8) → 0 T (9). Indeed, the hysteresis of scattering curves is observed also in our USANS experiments. In particular, it is clearly seen that the three curves (1, 5 and 9) obtained in the absence of a magnetic field are well separated from the family of curves obtained in the presence of a magnetic field. Moreover, these three curves do not coincide and they are progressively shifted towards the curves obtained in the presence of a magnetic field displaying the "memory" effect. Further, it is observed that the scattering curves 2 and 6 at 0.2 T corresponding to the ascending part of the hysteresis curve (i.e., when the field was increased from 0 to 0.2 T) are distinctly different from the scattering curves 4 and 8 at 0.2 T when the field was decreased from 0.35 T to 0.2 T and at 0.35 T. Interestingly, the scattering curves 4 and 8 seem to be similar to the curves 3 and 7 obtained in the maximum magnetic field.

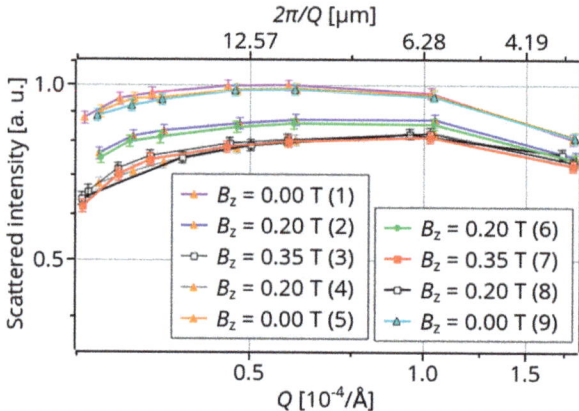

Figure 5. USANS curves of the MAE sample with 80 mass% of Fe and a soft matrix for different magnetic fields.

To qualitatively explain the observed differences in scattered curves, let us recall the known results on SANS scattering in a magnetic fluid [31]. The angle between the Q-vector and the magnetic field direction was set to be 0° (Q-vector parallel to the magnetic field direction), 30° and 60°. In a magnetic fluid, the magnetic particles formed chain-like structures and a pronounced (Bragg) peak, corresponding to a resulting particle-particle distance was observed. The position of this peak was practically independent of the angle between the magnetic field vector and the Q-vector. From this one may conclude that for magnetic fluid, the inter-particle distance in chains determines the position of the scattering peak. In MAEs, the displacements of particles are limited by the matrix elasticity. For high concentrations of magnetizable particles in MAEs, numerical simulations showed that formation of chain-like structures may become impossible due to purely geometrical constraints [32]. Therefore, in MAEs it is more reasonable to speak about formation of elongated particle aggregates and not about chain formation. A physical picture of changes in internal MAE microstructure was proposed in [33]. Obviously, the arrangement of magnetic particle in an MAE becomes anisotropic. The corresponding physical properties, e.g., the (dielectric) permittivity become anisotropic as well [34]. We were able to observe the anisotropy of neutron scattering in MAE samples using a very-small-angle neutron scattering (VSANS) instrument, but there we could not avoid multiple scattering [35].

In our presented USANS experiments, the Q-vector is perpendicular to the magnetic field direction. We may speculate that typically structures larger than 25 µm (elongated clusters) are formed in a highly filled sample (80 mass% of iron, see Figure 5). For such a highly loaded MAE, the existence of the three-dimensional magnetic–filler network already in the absence of a magnetic field can be expected. Intuitively, one would anticipate that the structures would elongate in field direction and, thus, this dimension would grow, while perpendicular structures would rather shrink by such alignment. The presented data can be interpreted as a reduction of structure's size in the direction perpendicular to the field direction, and thus loss of scattering at lower Q. In Figure 5, the formation of a local maximum at $Q \approx 10^{-4}$ Å^{-1} corresponding to the typical size of about 6 µm is observed, which is close to the mean particle diameter. This would suggest the formation of elongated aggregates of aligned individual particles.

However, such explanation does not work well for Figures 2 and 3 where a tendency to formation of a local maximum corresponding to the typical size of roughly 20 µm or larger can be assumed after what was said above. Also for these samples, loss of scattering for lower Q is observed. The typical size of 20 µm would be larger than the distance between uniformly distributed spherical particles with the mean diameter as used in our samples, corresponding to several particle diameters. Clearly, the particle concentration in Figures 2 and 3 is much smaller than in Figure 5. It is unlikely that large particle aggregates formed in these samples in the non-magnetized state. The origin of the observed effects may be in the resulting anisotropy of neutron scattering due to magnetically induced anisotropy of particle distribution. Such an effect has been observed in [31] where it has been found that the maximum amplitude of scattering is in the direction parallel to the field direction.

The hysteresis of Figure 5 becomes more clearly visible in Figure 6 where we plotted the area under the interpolated scattering curve in dependence on the magnetic flux density. The points in Figure 6 have been obtained in the following way. Each curve in Figure 6 was linearly interpolated between the measurement points and then integrated, the result of integration was divided by the length of the Q range and normalized by the maximum of the available results.

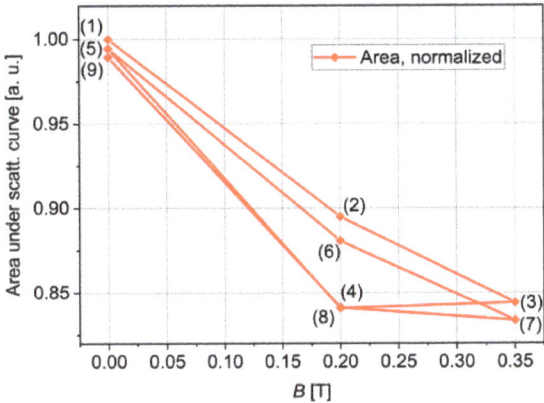

Figure 6. Normalized area under interpolated USANS scattering curve versus the applied magnetic field.

The hysteresis in Figure 6 is attributed to the magnetic-field-induced restructuring of filler particles, which is also the origin of magnetic hysteresis in such a material. It is known that the CIP particles are soft magnetic and possess a negligible magnetic hysteresis loop at room temperature [36]. In soft MAEs, the magnetic hysteresis appears in a particular form of a "pinched" hysteresis loop and it completely disappears at low temperatures when the matrix becomes stiff [37]. The relation of hysteresis of magnetic properties to the hysteresis of the consolidation of filler particles into elongated aggregates was explained by Zubarev et al. [30].

4. Conclusions and Outlook

Ultra-small-angle neutron scattering is a sensitive method for characterization of magnetoactive elastomers with micrometer-sized inclusions. We were able to observe the magnetic-field-induced changes in the scattering curves of soft MAEs in moderate magnetic fields of 200 and 350 mT. We believe that these changes are mainly due to modifications in the mutual arrangement of magnetic particles in a soft polymer matrix. It can be expected that the effect will be more pronounced in larger magnetic fields (say 600 mT), which should be verified in the future using another experimental setup. The USANS measurements confirm the previously put forward hypothesis that the hysteresis of physical properties of MAEs is related to the restructuring of filler particles.

To investigate the anisotropy of the field-direction-dependent (due to structural changes) nuclear neutron scattering and the contribution of magnetic neutron scattering, the experiments should be performed for different angles between the Q-vector and an applied magnetic field. This means that an experimental setup should allow for rotation of an electromagnet in the plane, which is perpendicular to the neutron beam. The usage of polarized neutron beams may be also helpful.

USANS experiments at a low temperature (<150 K) would allow one to immobilize the filler particles in an elastomer matrix [37] and investigate the neutron scattering on "frozen" MAE samples in the absence of an external magnetic field.

The interpretation of the scattering curves remains an open question for future research. A suitable theory has to be developed for describing the scattering from the elongated agglomerates of micro-particles. We believe that the analysis of USANS on MAEs may shed light on the processes of restructuring of ferromagnetic filler particles in a magnetic field. Usage of smaller filler particles (≈ 1 μm) may help to utilize the available Q-range more effectively and clarify the origin of the local maximum in the scattering curves.

Author Contributions: Conceptualization, M.S.; methodology, H.L. and J.K.; validation, J.K. and H.L.; formal analysis, J.K., H.L., M.S. and I.A.B.; investigation, H.L. and M.S.; resources, I.A.B. and H.L.; data curation, J.K. and H.L.; writing—original draft preparation, M.S. and J.K.; writing—review and editing, M.S., J.K. and H.L.; visualization, J.K. and I.A.B.; funding acquisition, M.S. All authors have read and agreed to the published version of the manuscript.

Funding: The research of I.A.B. was funded by the State Conference of Women and Equality Officers at Bavarian Universities (LaKoF Bavaria, Ph.D. scholarship). Open Access Funding by the University of Vienna.

Data Availability Statement: The data that support the findings of this study are openly available at: https://doi.ill.fr/10.5291/ILL-DATA.1-04-130 (accessed on 12 May 2021; registration of ILL user account required).

Acknowledgments: The authors thank Elena Kramarenko for useful discussions and Tobias Probst for assistance with preparation of experiments in Regensburg. The authors are grateful to Alexander Brunhuber and Wolfgang Kettl for providing MAE samples with stiff elastomer matrix. The authors thankfully acknowledge the anonymous reviewers for constructive criticism and useful suggestions.

Conflicts of Interest: The authors declare no conflict of interest.

References

1. Filipcsei, G.; Csetneki, I.; Szilágyi, A.; Zrínyi, M. Magnetic Field-Responsive Smart Polymer Composites. In *Oligomers–Polymer Composites-Molecular Imprinting*; Gong, B., Sanford, A.R., Ferguson, J.S., Eds.; Advances in Polymer Science; Springer: Berlin/Heidelberg, Germany, 2007; pp. 137–189. ISBN 978-3-540-46830-1.
2. Li, Y.; Li, J.; Li, W.; Du, H. A State-of-the-Art Review on Magnetorheological Elastomer Devices. *Smart Mater. Struct.* **2014**, *23*, 123001. [CrossRef]
3. Ubaidillah; Sutrisno, J.; Purwanto, A.; Mazlan, S.A. Recent Progress on Magnetorheological Solids: Materials, Fabrication, Testing, and Applications. *Adv. Eng. Mater.* **2015**, *17*, 563–597. [CrossRef]
4. Menzel, A.M. Tuned, Driven, and Active Soft Matter. *Phys. Rep.* **2015**, *554*, 1–45. [CrossRef]
5. Odenbach, S. Microstructure and Rheology of Magnetic Hybrid Materials. *Arch. Appl. Mech.* **2016**, *86*, 269–279. [CrossRef]
6. López-López, M.; Durán, J.; Iskakova, L.; Zubarev, A. Mechanics of Magnetopolymer Composites: A Review. *J. Nanofluids* **2016**, *5*, 479–495. [CrossRef]
7. Weeber, R.; Hermes, M.; Schmidt, A.M.; Holm, C. Polymer Architecture of Magnetic Gels: A Review. *J. Phys. Condens. Matter* **2018**, *30*, 063002. [CrossRef]
8. Shamonin, M.; Kramarenko, E.Y. Chapter 7–Highly Responsive Magnetoactive Elastomers. In *Novel Magnetic Nanostructures*; Domracheva, N., Caporali, M., Rentschler, E., Eds.; Advanced Nanomaterials; Elsevier: Amsterdam, The Netherlands, 2018; pp. 221–245. ISBN 978-0-12-813594-5.
9. Bastola, A.K.; Paudel, M.; Li, L.; Li, W. Recent Progress of Magnetorheological Elastomers: A Review. *Smart Mater. Struct.* **2020**, *29*, 123002. [CrossRef]
10. Cantera, M.A.; Behrooz, M.; Gibson, R.F.; Gordaninejad, F. Modeling of Magneto-Mechanical Response of Magnetorheological Elastomers (MRE) and MRE-Based Systems: A Review. *Smart Mater. Struct.* **2017**, *26*, 023001. [CrossRef]
11. Abramchuk, S.; Kramarenko, E.; Stepanov, G.; Nikitin, L.V.; Filipcsei, G.; Khokhlov, A.R.; Zrínyi, M. Novel Highly Elastic Magnetic Materials for Dampers and Seals: Part I. Preparation and Characterization of the Elastic Materials. *Polym. Adv. Technol.* **2007**, *18*, 883–890. [CrossRef]
12. Borbáth, T.; Günther, S.; Borin, D.; Gundermann, T.; Odenbach, S. XμCT Analysis of Magnetic Field-Induced Phase Transitions in Magnetorheological Elastomers. *Smart Mater. Struct.* **2012**, *21*. [CrossRef]
13. An, H.-N.; Picken, S.J.; Mendes, E. Direct Observation of Particle Rearrangement during Cyclic Stress Hardening of Magnetorheological Gels. *Soft Matter* **2012**, *8*, 11995–12001. [CrossRef]

14. Schümann, M.; Odenbach, S. In-Situ Observation of the Particle Microstructure of Magnetorheological Elastomers in Presence of Mechanical Strain and Magnetic Fields. *J. Magn. Magn. Mater.* **2017**, *441*, 88–92. [CrossRef]
15. Gundermann, T.; Cremer, P.; Löwen, H.; Menzel, A.M.; Odenbach, S. Statistical Analysis of Magnetically Soft Particles in Magnetorheological Elastomers. *Smart Mater. Struct.* **2017**, *26*, 045012. [CrossRef]
16. Avdeev, M.V.; Aksenov, V.L. Small-Angle Neutron Scattering in Structure Research of Magnetic Fluids. *Phys. Uspekhi* **2010**, *53*, 971. [CrossRef]
17. Odenbach, S.; Schwahn, D.; Stierstadt, K. Evidence for Diffusion-Induced Convection in Ferrofluids from Small-Angle Neutron Scattering. *Z. Für Phys. B Condens. Matter* **1995**, *96*, 567–569. [CrossRef]
18. Aksenov, V.; Avdeev, M.; Balasoiu, M.; Rosta, L.; Török, G.; Vekas, L.; Bica, D.; Garamus, V.; Kohlbrecher, J. SANS Study of Concentration Effect in Magnetite/Oleic Acid/Benzene Ferrofluid. *Appl. Phys. A* **2002**, *74*, s943–s944. [CrossRef]
19. Pop, L.M.; Hilljegerdes, J.; Odenbach, S.; Wiedenmann, A. The Microstructure of Ferrofluids and Their Rheological Properties. *Appl. Organomet. Chem.* **2004**, *18*, 523–528. [CrossRef]
20. Balasoiu, M.; Craus, M.L.; Plestil, J.; Haramus, V.; Erhan, R.; Lozovan, M.; Kuklin, A.I.; Bica, I. Microstructure of Magnetite Doped Elastomers Investigated by SAXS and SANS. 2008, 11. Available online: https://www.researchgate.net/publication/242829586_Microstructure_of_magnetite_doped_elastomers_investigated_by_SAXS_and_SANS (accessed on 12 May 2021).
21. Balasoiu, M.; Craus, M.L.; Anitas, E.M.; Bica, I.; Plestil, J.; Kuklin, A.I. Microstructure of Stomaflex Based Magnetic Elastomers. *Phys. Solid State* **2010**, *52*, 917–921. [CrossRef]
22. Balasoiu, M.; Lebedev, V.T.; Orlova, D.N.; Bica, I.; Raikher, Y.L. SANS Investigation of a Ferrofluid Based Silicone Elastomer Microstructure. *J. Phys. Conf. Ser.* **2012**, *351*, 012014. [CrossRef]
23. Balasoiu, M.; Lebedev, V.T.; Raikher, Y.L.; Bica, I.; Bunoiu, M. The Implicit Effect of Texturizing Field on the Elastic Properties of Magnetic Elastomers Revealed by SANS. *J. Magn. Magn. Mater.* **2017**, *431*, 126–129. [CrossRef]
24. Pyanzina, E.S.; Sánchez, P.A.; Cerdà, J.J.; Sintes, T.; Kantorovich, S.S. Scattering Properties and Internal Structure of Magnetic Filament Brushes. *Soft Matter* **2017**, *13*, 2590–2602. [CrossRef]
25. Borin, D.Y.; Bergmann, C.; Odenbach, S. Characterization of a Magnetic Fluid Exposed to a Shear Flow and External Magnetic Field Using Small Angle Laser Scattering. *J. Magn. Magn. Mater.* **2020**, *497*, 165959. [CrossRef]
26. Zákutná, D.; Graef, K.; Dresen, D.; Porcar, L.; Honecker, D.; Disch, S. In Situ Magnetorheological SANS Setup at Institut Laue-Langevin. *Colloid Polym. Sci.* **2021**, *299*, 281–288. [CrossRef]
27. Sorokin, V.V.; Belyaeva, I.A.; Shamonin, M.; Kramarenko, E.Y. Magnetorheological Response of Highly Filled Magnetoactive Elastomers from Perspective of Mechanical Energy Density: Fractal Aggregates above the Nanometer Scale? *Phys. Rev. E* **2017**, *95*, 062501. [CrossRef] [PubMed]
28. Belyaeva, I.A.; Kramarenko, E.Y.; Shamonin, M. Magnetodielectric Effect in Magnetoactive Elastomers: Transient Response and Hysteresis. *Polymer* **2017**, *127*, 119–128. [CrossRef]
29. Hainbuchner, M.; Villa, M.; Kroupa, G.; Bruckner, G.; Baron, M.; Amenitsch, H.; Seidl, E.; Rauch, H. The New High Resolution Ultra Small-Angle Neutron Scattering Instrument at the High Flux Reactor in Grenoble. *J. Appl. Crystallogr.* **2000**, *33*, 851–854. [CrossRef]
30. Zubarev, A.; Chirikov, D.; Borin, D.; Stepanov, G. Hysteresis of the Magnetic Properties of Soft Magnetic Gels. *Soft Matter* **2016**, *12*. [CrossRef]
31. Barrett, M.; Deschner, A.; Embs, J.P.; Rheinstädter, M.C. Chain Formation in a Magnetic Fluid under the Influence of Strong External Magnetic Fields Studied by Small Angle Neutron Scattering. *Soft Matter* **2011**, *7*, 6678–6683. [CrossRef]
32. Romeis, D.; Toshchevikov, V.; Saphiannikova, M. Elongated Micro-Structures in Magneto-Sensitive Elastomers: A Dipolar Mean Field Model. *Soft Matter* **2016**, *12*, 9364–9376. [CrossRef] [PubMed]
33. Snarskii, A.A.; Zorinets, D.; Shamonin, M.; Kalita, V.M. Theoretical Method for Calculation of Effective Properties of Composite Materials with Reconfigurable Microstructure: Electric and Magnetic Phenomena. *Phys. A Stat. Mech. Appl.* **2019**, *535*, 122467. [CrossRef]
34. Snarskii, A.A.; Shamonin, M.; Yuskevich, P.; Saveliev, D.V.; Belyaeva, I.A. Induced anisotropy in composite materials with reconfigurable microstructure: Effective medium model with movable percolation threshold. *Phys. A Stat. Mech. Appl.* **2020**, *560*, 125170. [CrossRef]
35. Pipich, V. Magnetic-Field-Induced Structural Changes in Compliant Magnetorheological Elastomers. Private Communication, 2018.
36. Stepanov, G.V.; Borin, D.Y.; Raikher, Y.L.; Melenev, P.V.; Perov, N.S. Motion of Ferroparticles Inside the Polymeric Matrix in Magnetoactive Elastomers. *J. Phys. Condens. Matter* **2008**, *20*, 204121. [CrossRef] [PubMed]
37. Bodnaruk, A.V.; Brunhuber, A.; Kalita, V.M.; Kulyk, M.M.; Kurzweil, P.; Snarskii, A.A.; Lozenko, A.F.; Ryabchenko, S.M.; Shamonin, M. Magnetic Anisotropy in Magnetoactive Elastomers, Enabled by Matrix Elasticity. *Polymer* **2019**, *162*, 63–72. [CrossRef]

Article

The Analysis of Periodic Order in Monolayers of Colloidal Superballs

Daniël N. ten Napel [1], Janne-Mieke Meijer [2] and Andrei V. Petukhov [1,3,*]

[1] Van 't Hoff Laboratory for Physical and Colloid Chemistry, Debye Institute for Nanomaterials Science, Utrecht University, Padualaan 8, 3584 CH Utrecht, The Netherlands; d.n.tennapel@uu.nl

[2] Soft Matter and Biological Physics, Department of Applied Physics, Eindhoven University of Technology, P.O. Box 513, 5600 MB Eindhoven, The Netherlands; j.m.meijer@tue.nl

[3] Laboratory of Physical Chemistry, Department of Chemical Engineering and Chemistry, Eindhoven University of Technology, P.O. Box 513, 5600 MB Eindhoven, The Netherlands

* Correspondence: a.v.petukhov@uu.nl; Tel.: +31-30-253-1167

Abstract: The characterization of periodic order in assemblies of colloidal particles can be complicated by the coincidence of Bragg diffraction peaks of the structure and minima in the form factor of the particles. Here, we demonstrate a general strategy to overcome this problem that is applicable to all low-dimensional structures. This approach is demonstrated in the application of small-angle X-ray scattering (SAXS) for the characterization of monolayers of colloidal silica superballs prepared using the unidirectional rubbing method. In this method, the ordering of the colloidal superballs is achieved by mechanically rubbing them onto a polydimethylsiloxane (PDMS)-coated surface. Using three differently shaped superballs, ranging from spherical to almost cubic, we show that certain Bragg peaks may not appear in the diffraction patterns due to the presence of minima in the form factor. We show that these missing Bragg peaks can be visualized by imaging the colloidal monolayers at various orientations. Moreover, we argue that the same strategy can be applied to other techniques, such as neutron scattering.

Keywords: colloidal superballs; colloidal monolayers; SAXS; scattering; form factor

Citation: ten Napel, D.N.; Meijer, J.-M.; Petukhov, A.V. The Analysis of Periodic Order in Monolayers of Colloidal Superballs. *Appl. Sci.* **2021**, *11*, 5117. https://doi.org/10.3390/app11115117

Academic Editor: Sebastian Jaksch

Received: 10 May 2021
Accepted: 26 May 2021
Published: 31 May 2021

Publisher's Note: MDPI stays neutral with regard to jurisdictional claims in published maps and institutional affiliations.

Copyright: © 2020 by the authors. Licensee MDPI, Basel, Switzerland. This article is an open access article distributed under the terms and conditions of the Creative Commons Attribution (CC BY) license (https://creativecommons.org/licenses/by/4.0/).

1. Introduction

Colloidal monolayers are used in a wide variety of applications, such as anti-reflective coatings, photonic materials, photovoltaics and biosensors [1,2]. Alternatively, the assembly of colloids into monolayers is often used as a step in colloidal syntheses in order to locally modify the particles in order, for example, to produce Janus colloids [3–5]. Various methods are known for the preparation of colloidal monolayers, such as drying methods, spin coating, the assembly on liquid interfaces and the horizontal and vertical deposition methods [2]. Each of these preparation methods have their own strengths and weaknesses. It was only recently shown, however, that colloidal monolayers could also be prepared by mechanically rubbing dried colloids between two sticky surfaces such as polydimethylsiloxane (PDMS) [6,7]. This quick, simple and inexpensive method produces high-quality colloidal monolayers where the colloidal crystal grains align in the direction in which the mechanical rubbing was applied.

So far, the unidirectional rubbing method has only been used for the monolayer formation of spherical particles. However, the shape of the colloidal particles can greatly influence the structures that are formed [8]. By employing the unidirectional rubbing method for the assembly of anisotropic colloids, colloidal monolayers that possess unique properties may be formed, which might be of interest in one of the many colloidal monolayer applications. Of specific interest are superball-shaped colloids, i.e., shapes varying from spheres via cubes with rounded edges to perfect cubes, because their variable shape allows smooth phase transitions to occur [9]. For instance, upon allowing superball-shaped colloids to sediment, plastic crystals and solid–solid phase transitions were observed [10,11], whereas upon assembling these particles into monolayers using the horizontal and vertical deposition method both predicted densest packings were found as well as a continuous transition between the two lattice structures [12,13]. Furthermore, upon the addition of depletants, the resulting crystal phases were found to depend on both the shape of the superballs and the radius of gyration of the added depletant relative to the size of the superballs [9,14]. By employing the unidirectional rubbing method for the organization of colloidal superballs into monolayers it can be expected that monolayers with unique organizations can be achieved as a result of the directional force applied during its assembly.

The periodic structures in colloidal monolayers can be studied, on the one hand, using microscopy techniques, which are superior in revealing local densely packed structures. On the other hand, to characterize long-range ordering on the colloidal scale, small-angle neutron scattering (SANS) [15] and small-angle X-ray scattering (SAXS) [16] are often applied. Although these two techniques are very similar in their general approach and theoretical models, there are many factors that determine which of these scattering techniques is most suitable for a certain study. For example, while for X-rays the scattering contrast is essentially determined by the total local electron density, which is approximately proportional to the mass density, neutron scattering contrast has much higher sensitivity to most lighter atoms with fewer electrons. Moreover, SANS offers great opportunities for contrast variation by using isotopes, most remarkably hydrogen/deuterium. This allows the revelation of minor structural details [17] or the minimization of multiple scattering [18]. In addition, the scattering of spin-polarized neutrons is sensitive to the magnetic ordering in the sample [19]. Wide opportunities can be opened by combining SANS with advanced spin-echo techniques, such as spin-echo small-angle neutron scattering (SESANS), which can access large-period structures [20] as well as nanostructures along with their nanosecond temporal dynamics [21,22]. The recent developments in synchrotron sources, X-ray detectors, and X-ray optics paved the way to millisecond SAXS data acquisition, microradian angular resolution, the possibility to scan samples with a submicrometer-sized beam and the application of coherent X-ray scattering techniques [23,24]. Due to all these achievements, contemporary SAXS and SANS became truly complementary techniques.

In this work we have studied the ordered monolayer structures of colloidal superballs similar in size but with different shapes, ranging from spherical to almost cubic, prepared using the unidirectional rubbing method. Since these structures are static, the contrast is high, and because high angular resolution is required due to the large structure period, SAXS is the technique of choice in the present work. The small-angle scattering from a periodic arrangement of particles does not only depend on their mutual spatial arrangement, but also on the shape of the individual particles. The latter is described by the form factor. For spherically symmetric particles the form factor is isotropic because the X-rays are scattered equally in all directions and the form factor can be calculated analytically. For non-spherical particle shapes, however, the form factor is no longer isotropic, which complicates the analysis of the scattering patterns. It was previously shown by Meijer et al. that the form factor of hollow silica superballs with a thin shell and distinct cubic shape become cross-like with little intensity in the corners of the diffraction patterns [11]. These effects in the form factor will result in Bragg peaks in these locations to be less visible, which complicates the structural analysis. As we faced in the present study, the presence of minima in the form factor of hollow superball particles can make many important diffraction peaks invisible, which made the structure characterization complicated if not impossible.

Here, we present a strategy for the measurement and analysis of the SAXS diffraction patterns of monolayers of colloidal silica superballs prepared using the unidirectional rubbing method. We show that by performing the measurements under a range of rotations, a detailed analysis can be performed of both the form factor and structure factor. In this case the q-values of the Bragg peaks can be shifted away from the minima in the form factor. The effect of sample rotation is first illustrated using spheres for which the form factor is isotropic and can be calculated analytically. We then show for the more cubic superballs, for which the form factor cannot be calculated analytically, that this rotation allows us to visualize and assign all Bragg diffraction peaks for a full structural analysis. Finally, we present a model for the phase transition observed in the lattices of colloidal superball monolayers prepared using the unidirectional rubbing method upon varying the shape of the colloidal superballs from spherical to almost cubic.

2. Theory

In SAXS the scattered intensity is the product of the form factor and the structure factor:

$$I(\vec{q}) = P(\vec{q})S(\vec{q}), \qquad (1)$$

where $\vec{q}$ is the scattering vector. The form factor $P(\vec{q})$ is determined by the shape and size of the individual particles. For uniform spherical particles the form factor is radially symmetric and equal to:

$$P(\vec{q}) = P(q) = \frac{3[\sin(qR) - (qR)\cos(qR)]}{(qR)^3}, \qquad (2)$$

where q is the length of the scattering vector and R is the radius of the spherical particle. Due to the symmetry of the spherical shape, the form factor does not depend on the particle orientation. For non-spherical particles, however, the form factor does depend on the orientation of the particle. While most anisotropic shapes do not have an analytical solution, their form factor can always be calculated numerically.

The structure factor $S(\vec{q})$, on the other hand, is determined by the structural assembly of the particles. Consider an ideal two-dimensional lattice with lattice points at distance vectors of:

$$\vec{r} = m\vec{a}_1 + n\vec{a}_2, \qquad (3)$$

where m and n are integers and $\vec{a}_1$ and $\vec{a}_2$ are the primitive lattice vectors. The structure factor of any structural assembly is equal to:

$$S(\vec{q}) = \frac{1}{N}\left|\sum_{n=1}^{N} e^{i\vec{q}\cdot\vec{r}_n}\right|^2, \qquad (4)$$

where N is the total number of particles and $\vec{r}_n$ is the distance vector of the nth particle. For an ideal two-dimensional lattice the structure factor is non-zero at:

$$\vec{q} = h\vec{b}_1 + k\vec{b}_2, \qquad (5)$$

for any value of q_z. Here, h and k are the Miller indices, $\vec{b}_1$ and $\vec{b}_2$ are the primitive reciprocal basis vectors and q_z is the component of the scattering vector that is parallel to the incoming X-ray beam. The angle between these primitive reciprocal basis vectors is defined as β. The primitive reciprocal basis vectors are related to the primitive lattice vectors as follows:

$$\vec{b}_1 = 2\pi\frac{\vec{a}_2 \times \vec{n}}{|\vec{a}_1 \times \vec{a}_2|}, \vec{b}_2 = 2\pi\frac{\vec{n} \times \vec{a}_1}{|\vec{a}_1 \times \vec{a}_2|}, \qquad (6)$$

where $\vec{n}$ is a vector normal to the two-dimensional lattice plane. Finally, the angle between the primitive lattice vectors is $\alpha = 180° - \beta$.

3. Materials and Methods

3.1. Preparation of the Colloidal Superball Monolayers

The solid colloidal silica spheres were synthesized using the Stöber process [25]. Two types of cube-like hollow colloidal superballs were synthesized by coating precursor colloidal hematite cubes with a layer of silica and subsequently dissolving the magnetic core particle [9,26]. The superball family describes the shape that smoothly interpolates between a sphere and a cube. The contours of the superball shape are defined by the following formula [13]:

$$\left|\frac{2x}{D}\right|^m + \left|\frac{2y}{D}\right|^m + \left|\frac{2z}{D}\right|^m = 1, \qquad (7)$$

where m is the shape parameter, D the diameter (face-to-face length) of the superballs and x, y and z are the coordinates of the surface along the x-, y- and z-axes, respectively. The average shape parameter m and diameter D of the synthesized superball colloids were determined from transmission electron microscopy (TEM) images using a custom-written analysis script written in Python. The resulting spherical, semi-cubic and cube-like particles were found to possess average shape parameters of $m = 2$, 2.87 and 3.63 and diameters of $D = 1.10$, 0.96 and 1.01 µm, respectively. These colloidal silica superballs were assembled into monolayers by mechanically rubbing them onto polydimethylsiloxane (PDMS)-coated glass slides (#0, 60 mm × 24 mm × 0.085 − 0.12 mm) using a custom-made device. For all samples, this device was set to run 10 rubbing cycles consisting of a back and forth motion of a PDMS cylinder with a diameter of 1 cm at a rubbing pressure of 200 g and a rubbing speed of 3 mm/s. At the end of each rubbing motion, the PDMS cylinder was lifted over the remaining colloidal powder. Finally, the excess colloidal powder was removed using a strong nitrogen flow. TEM images of the individual colloidal superballs and scanning electron microscopy (SEM) images of their corresponding monolayers are depicted in Figure 1a–f.

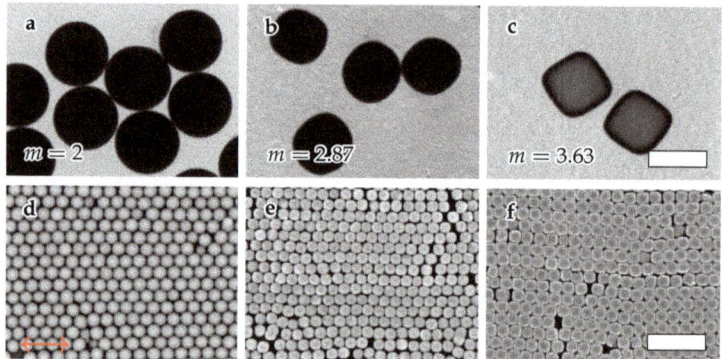

Figure 1. Representative (**a**–**c**) transmission electron microscopy (TEM) images of the colloidal superballs and (**d**–**f**) scanning electron microscopy (SEM) images of their corresponding monolayers assembled using the unidirectional rubbing method. The direction of the mechanical rubbing is indicated in the images using a red arrow. The scale bars represent (**a**–**c**) 1 and (**d**–**f**) 5 µm.

3.2. Analysis of the Colloidal Superball Monolayers Using SAXS

The SAXS measurements were performed at the BM26B DUBBLE beamline of the ESRF in Grenoble, France using the microradian setup [23] employing compound refractive lenses [27]. The X-ray beam with a photon energy of 12 kV was focused on a CCD with a resolution of 4008 × 2671 pixels and pixel size of 22 × 22 µm (Photonic Science) placed at a distance of 7.5 m from the sample. A wedge-shaped beam-stop was used in order to prevent damage to the detector that can result from an exposure to the direct X-ray beam. The colloidal monolayers were placed perpendicular to the X-ray beam with the rubbing direction along to the horizontal plane (x-axis). The samples were rotated around the x-axis from $\phi = -60$ to $60°$. The SAXS setup is schematically depicted in Figure 2.

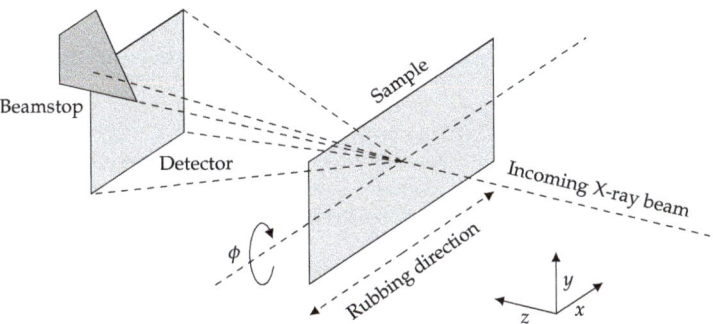

Figure 2. A schematic representation of the small-angle X-ray scattering (SAXS) setup. Here, the monolayers of colloidal superballs are rotated around the x-axis. The angle of rotation around this axis is denoted by ϕ. The incoming X-ray beam travels along the z-axis. The scattering patterns are imaged in x,y-space. A beam-stop was used to prevent damage to the detector. Note that this schematic representation is not drawn to scale.

2D SAXS patterns were obtained at different rotations of ϕ with an exposure time of 20 s. Dark current and background corrections were performed on the patterns before their analysis. The background images were obtained from a blank PDMS-coated glass slide at an equal rotation of ϕ. Every measured diffraction pattern image was subtracted with the background imaged at an equal rotation in the x-axis to correctly remove the scattering background that is caused by the PDMS-coated glass slide. Custom analysis scripts written in Python were used to determine the positions of the peaks in the cross sections of the diffraction pattern images and to numerically calculate the form factor of the different superball shapes. To model the form factor, we constructed the superballs, whether solid or hollow, on a three-dimensional raster with 30 points along each dimension. The interference between the points within the particles' shape was then calculated and converted via a Fourier transform.

4. Results and Discussion

To elucidate the role of the particles' form factor on the structural analysis in (X-ray) diffraction patterns, monolayers of superball colloids with three different shapes, namely spherical, semi-cubic and cube-like, were prepared using the unidirectional rubbing method and investigated with SAXS. These colloidal monolayers were rotated around the rubbing direction (x-axis). First, we focus on 2D SAXS patterns and their analysis for a monolayer obtained from spherical colloids that can be described as a superball with a shape parameter of $m = 2$. Figures 3a–c show the SAXS patterns obtained for a monolayer prepared from solid colloidal silica spheres with $D = 1.10\,\mu\text{m} \pm 3.2\%$. The three typical diffraction patterns imaged at rotations of $\phi = 0$, 20 and 40° all show distinct peaks that indicate the spheres have formed well-defined periodic structures in the monolayer. The presence of separate peaks, instead of a series of co-axial rings, indicate that the crystal orientation within the imaged area of the colloidal monolayer is fixed. This is a known result of the unidirectional rubbing method where the crystal grains align in the direction of rubbing [6,7].

Figure 3a shows the diffraction pattern imaged at a rotation of $\phi = 0°$, where six-fold Bragg diffraction peaks indicate that the crystal structure of the colloidal monolayer is a two-dimensional close-packed hexagonal lattice. The peaks centered along the vertical direction of the images can be identified as the $hk = 10$ and 20 peaks of a hexagonal lattice, as indicated in Figure 3a. Upon rotating the sample around the x-axis, the Bragg peaks move in the vertical y-direction, which makes the diffraction pattern appear elongated (Figure 3b,c). In addition, it can be seen that the intensities of the peaks in the diffraction patterns change relative to each other. Specifically, the 10 peak is strongly visible in the pattern at $\phi = 0$ and 20°, but disappears at $\phi = 40°$, while the 20 peak is visible at $\phi = 0$

and 40°, but disappears at $\phi = 20°$. By contrast, the 30 peak is only slightly visible at $\phi = 20°$ and invisible at both $\phi = 0$ and 40°. These trends can be more easily visualized using the cross sections of the scattering patterns in the vertical direction at $q_x = 0$ µm. Our analysis revealed that upon rotating the sample the positions of the maxima in the structures shifted to a higher q_y-values. This shift is directly related to the rotation and by scaling the scattering vector q_y with a factor of $\cos(\phi)$ all peaks in the various cross sections fell at equal values of $q_y \cos(\phi)$, as shown in Figure 3d for $\phi = 0, 10, 20, 30$ and 40°. In this case we find maxima at $q_y \cos(\phi) = 5.95, 11.90$ and $17.85\,\mu m^{-1}$ for the 10, 20 and 30 peaks, respectively. Thus, by considering the cross sections of the diffraction patterns imaged at various rotations of ϕ, a more precise analysis can be performed to locate the positions of the various hk peaks.

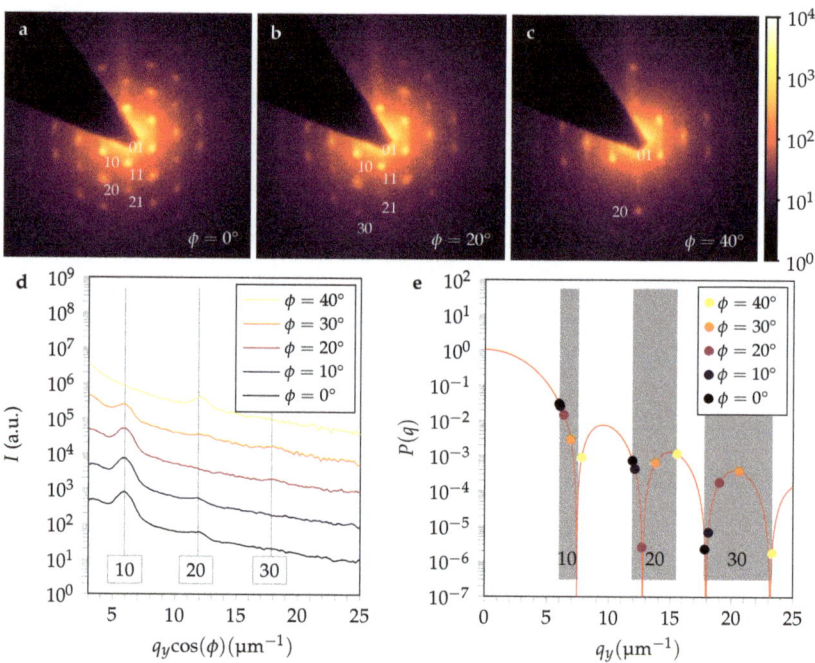

Figure 3. (**a**–**c**) 2D SAXS patterns of a monolayer prepared from colloidal silica spheres with an average diameter of $D = 1.10\,\mu m \pm 3.2\%$ imaged at rotations around the x-axis of ϕ. The Bragg peaks are labelled on their left. (**d**) Cross sections of the diffraction patterns at different ϕ taken at $q_x = 0\,\mu m^{-1}$ versus $q_y \cos(\phi)$, revealing that the positions of the identified hk peaks occur at specific $q_y \cos(\phi)$ positions. (**e**) Positions of the 10, 20 and 30 peaks for different ϕ overlayed on top of the calculated spherical form factor $P(q)$.

The strong changes in the intensity of the peaks observed upon rotating the sample are due to the presence of minima in the form factor. The distances of periodicity in the sample in the y-direction decreased upon rotation and thus correspond to higher q_y-values in reciprocal space. Figure 3e shows this process for the positions of the 10, 20 and 30 peaks at different ϕ plotted on top of the calculated spherical form factor $P(q)$. It can be seen that the positions of the hk peaks shifted to higher q_y-values with increasing ϕ, thus resulting in changes in their relative intensities. This example shows that if a large enough range in the scattering vector q_y can be probed to shift the positions of these hk peaks away from the $P(q)$ minima, all peaks can be identified correctly. Similar analyses can be performed to determine the positions of the other hk peaks present in the scattering pattern. The 01 peak, for example, is visible in each of the diffraction patterns depicted in Figure 3a–c, whereas both the 11 and 21 peaks are only visible up to a rotation of $\phi = 20°$ (Figure 3a,b).

Next, we discuss the diffraction patterns of a monolayer prepared from hollow colloidal silica superballs with $m = 2.87 \pm 13\%$ and $D = 0.96\,\mu m \pm 4.5\%$ that have a semicubic shape and, in addition, are now hollow shells, which will influence the form factor. Figure 4a–c show the 2D SAXS patterns for different rotations ϕ. Here, distinct Bragg peaks can again be observed, which we can identify as the different hk peaks as indicated in Figure 4a–c. Similar to the spherical colloids, we observed that different peaks are visible at different rotations. At $\phi = 0°$, the 10 peak is only barely visible while both the 20 and 30 peaks are clearly visible. Interestingly, and perhaps coincidentally, at $\phi = 20°$ no peaks are visible all along the vertical axis at $q_x = 0\,\mu m^{-1}$. At $\phi = 40°$, the 10 and 20 peaks appear again while the 30 peak is still invisible.

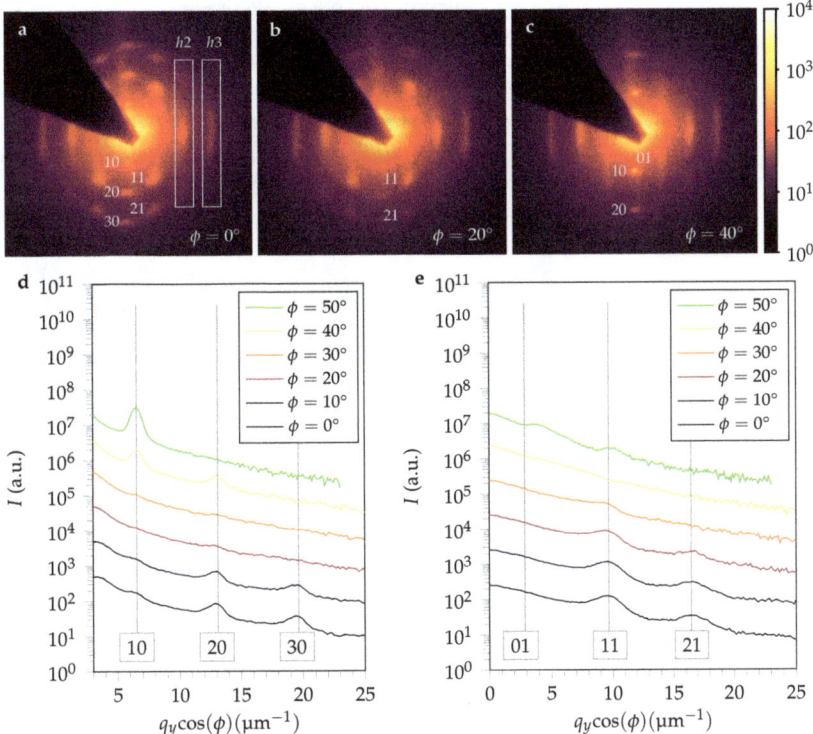

Figure 4. (**a**–**c**) 2D SAXS patterns of a monolayer prepared from hollow colloidal silica superballs with an average shape parameter of $m = 2.87 \pm 13\%$ and diameter of $D = 0.96\,\mu m \pm 4.5\%$ imaged at different rotations around the x-axis of ϕ. The Bragg peaks are labelled on their left. (**d**–**e**) Vertical cross sections of the diffraction patterns at (**d**) $q_x = 0\,\mu m^{-1}$ and (**e**) $q_x = 6.12\,\mu m^{-1}$. The vertical lines indicate the $q_y \cos(\phi)$ positions of the identified peaks.

Figure 4d shows the cross sections taken along the vertical axis at $q_x = 0\,\mu m^{-1}$. Again, in these cross sections the observed peak behavior can be more clearly visualized. The 10 peak is only visible at high ϕ values, while the 30 is only visible at low ϕ values and no peaks are observed at rotations in the x-axis of $\phi = 20$ and $30°$. The appearance and disappearance of the peaks in the scattering patterns upon rotation for these superball monolayers is distinctly different from the behavior observed for the monolayers consisting of spheres. The explanation of this behavior is that the particle form factor, now from a hollow and anisotropic particle, clearly influenced the appearance of structure factor peaks considerably, as expected. To further characterize and understand these non-trivial SAXS patterns, it can be useful to also consider vertical cross sections at different q_x positions. Figure 4e depicts the vertical cross sections taken at $q_x = 6.12\,\mu\text{m}^{-1}$, which contain the $h1$ peaks. Here, it can be seen that there is a clear evolution of the peaks. Specifically, the 01 peak becomes visible only at $\phi = 50°$. By contrast, the 11 and 21 peaks are both visible at up to $\phi = 20°$ and disappear at higher angles with only the 11 peak appearing again at $\phi = 50°$. The observation that both the 10 and the 01 peaks only appear upon a considerable rotation indicates that these peaks must be close to a minimum in the form factor. Furthermore, it can be seen that the $h2$ and $h3$ peaks appear to be broadened along the vertical y-direction, as indicated in Figure 4a–c.

The biggest influence of the particle form factor was observed for monolayers prepared from hollow superballs with $m = 3.63 \pm 12\%$ and $D = 1.01\,\mu\text{m} \pm 6.6\%$ that have a distinct cube-like shape. Figure 5a–c show the 2D SAXS patterns imaged at different rotations. Strikingly, here not even a single Bragg peak is significantly visible in the pattern at a rotation of $\phi = 0°$ as seen in Figure 5a. Upon rotating the sample to $\phi = 20$ and $40°$, some of the diffraction peaks can be made visible (Figure 5b,c).

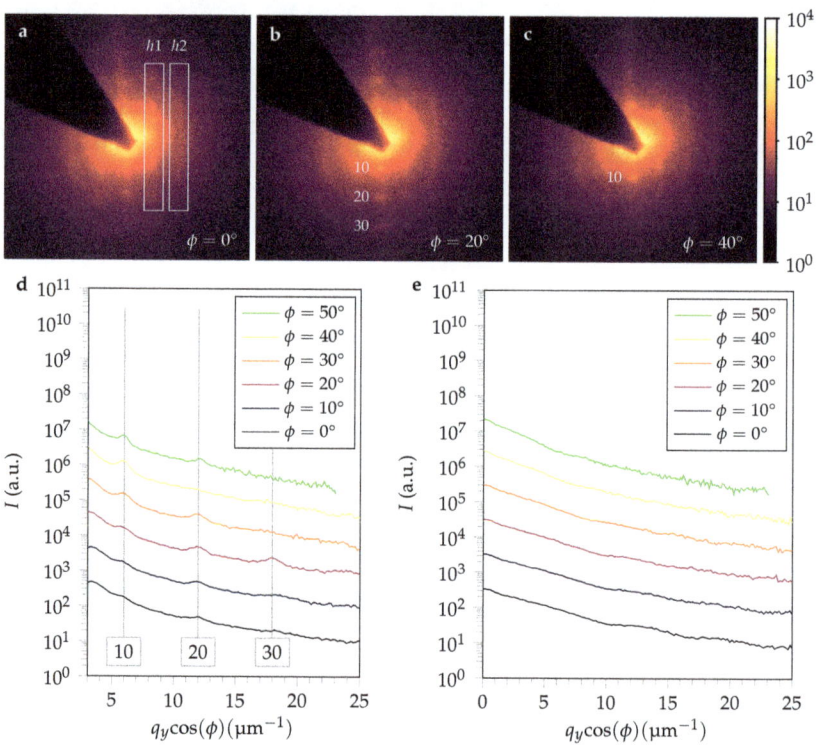

Figure 5. (**a–c**) 2D SAXS patterns of a monolayer prepared from hollow colloidal silica superballs with an average shape parameter of $m = 3.63 \pm 12\%$ and diameter of $D = 1.01\,\mu\text{m} \pm 6.6\%$ imaged at different rotations around the x-axis of ϕ. The Bragg peaks are labelled on their left. (**d,e**) Vertical cross sections of the diffraction patterns at (**d**) $q_x = 0\,\mu\text{m}^{-1}$ and (**e**) $q_x = 5.76\,\mu\text{m}^{-1}$. The vertical lines indicate the $q_y\cos(\phi)$ positions of the identified peaks.

Again, we examine the vertical cross sections of the diffraction patterns at $q_x = 0\,\mu\text{m}^{-1}$ (Figure 5d). From the evolution of the peaks, it is clear that the 10 peak appears in the patterns at $\phi = 30°$ and higher rotations, while the 20 and 30 peaks are only visible up to $\phi = 30°$. Furthermore, in the scattering patterns depicted in Figure 5a–c, some intensity can be seen where the $h1$ and $h2$ peaks are expected. These regions are highlighted in Figure 5a. In Figure 5e the cross sections of the diffraction patterns imaged at the various rotations ϕ taken at $q_x = 5.76\,\mu m^{-1}$ are depicted, corresponding to the intensity ascribed to the $h1$ peaks. Due to their broadness, however, no clear Bragg peaks can be assigned here as the $h1$ peaks appear to form a continuous band.

To further understand the appearance and disappearance of the peaks, we examine the different form factor contribution of the superball colloids used here, with spherical, semi-cubic and cube-like shapes. The form factor of a hollow superball is a function of the shape and size as well as the thickness of the shell for which there is no analytical solution. Therefore, we have calculated their form factors numerically. Figure 6a–c show the resulting calculated 2D form factors at a rotation of $\phi = 0°$, in which the semi-cubic and cube-like superballs lie flat on the PDMS-coated substrate, and an effectively 2D projection is obtained. Although the form factor for spherical particles has an analytical solution, as seen in Figure 3e, we included it here as a numerical calculation for a direct comparison. It can be seen that upon increasing the shape parameter m of the colloidal superballs, the isotropic rings visible for spheres ($m = 2$) began to show oscillations with maxima in the horizontal and vertical directions, i.e., in the directions of the flat sides of hollow superballs, and minima in the diagonal directions, i.e., the corners of the hollow superballs. These changes in the form factor explain why the structure factor peaks visible for the spherical particles disappear for the hollow superballs.

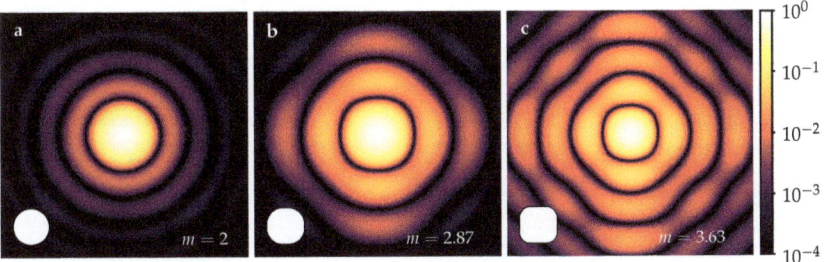

Figure 6. Numerically calculated form factors of the (**a**) solid spherical, (**b**) hollow semi-cubic and (**c**) hollow cube-like superballs at a rotation of $\phi = 0°$. The orientation of the superball particles is depicted in the insets of the images. Here, the intensity is limited to 4 orders of magnitude, which is approximately the same range the detector used in these experiments is able to measure.

Now that we have assigned all identifiable Bragg peaks in the diffraction patterns of the three superball shapes, we can begin to deduce their structure factors $S(\vec{q})$ and determine what these superball lattices look like in real space. From the positions of the 10, 20 and 30 peaks we can determine the length of the first primitive reciprocal basis vector, $\vec{b}_1$, according to Equation 5. For the spherical and semi-cubic superballs, we can determine the length of the second primitive reciprocal basis vector, $\vec{b}_2$, from the positions of the 01, 11 and 21 peaks. Since for the cube-like superballs the 01, 11, 21 peaks appear to form a continuous band, we can instead define $\vec{b}_{2x}$ as the horizontal component of $\vec{b}_2$. Here, $\vec{b}_{2x}$ is perpendicular to $\vec{b}_1$. The deduced structure factors for the spherical, semi-cubic and cube-like superballs along with the identified Bragg peaks and their primitive reciprocal basis vectors are depicted in Figure 7a–c, respectively. From the primitive reciprocal basis vectors $\vec{b}_1$ and $\vec{b}_2$, we can extract the real space primitive lattice vectors $\vec{a}_1$ and $\vec{a}_2$ and, in the case of the cubic-like superballs, $\vec{a}_{2x}$. Their corresponding real space lattices are shown in Figure 7d–f. For the spherical superballs we determined the length of the primitive lattice vectors to be $a_1 = 1.21 \pm 0.013\,\mu m$ and $a_2 = 1.19 \pm 0.004\,\mu m$, with an angle of $\alpha = 119.4 \pm 0.26°$. This confirms that these superballs indeed form a hexagonal lattice in real space with a well-defined and equal periodicity in each direction. For the semi-cubic superballs we found that $a_1 = 1.11 \pm 0.019\,\mu m$ and $a_2 = 1.04 \pm 0.011\,\mu m$, separated by an angle of $\alpha = 117.8 \pm 0.47°$. Here, since a_1 is significantly larger than a_2, these semi-cubic superballs form a distorted hexagonal (Λ_0) lattice as expected for the densest packing of superdisks in 2D [28]. From $a_1 > a_2$ we further conclude that these semi-cubic superballs are oriented with their flat faces in the horizontal and vertical directions. For the cubic-like superballs that form a more cubic-like lattice we find that $a_1 = 1.05 \pm 0.027$ µm and $a_{2x} = 1.08 \pm 0.044$ µm. Furthermore, from the presence of broad and continuous $h1$ and $h2$ bands in $S(\vec{q})$, we further conclude that these cube-like colloidal superballs form chains in the horizontal direction that are able to slide freely with respect to each other.

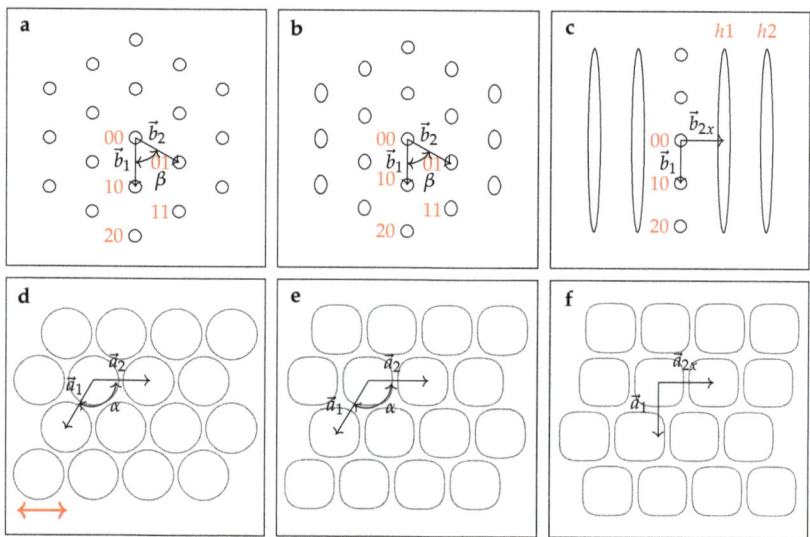

Figure 7. (a–c) The deduced structure factor $S(\vec{q})$ in reciprocal space and (d–f) the corresponding lattices in real space for the (a,d) solid spherical, (b,e) hollow semi-cubic and (c,f) hollow cube-like superballs. Here $\vec{b}_1$ and $\vec{b}_2$ are the primitive reciprocal basis vectors, separated by angle β, and $\vec{a}_1$ and $\vec{a}_2$, separated by angle α, are the corresponding primitive lattice vectors. For the cube-like superballs we have instead defined $\vec{b}_{2x}$ as the horizontal component that is perpendicular to $\vec{b}_1$. The Bragg peaks are labelled on their left. The direction of the mechanical rubbing is indicated in the images using a red arrow.

Thus, upon increasing the shape parameter m of the superballs, several structural changes are observed in their lattice structures. First, the superballs have formed chains in the horizontal direction in the monolayers, which are aligned to the direction of the mechanical rubbing. Upon increasing the shape parameter m, the ratio between a_1 and a_2 increases, which indicates that the spacing between the chains increases, as expected for superballs [28]. This leads to the ability for the chains to slide along each other for the most cubic-like superballs. Second, the broadening in the $h1$ and $h2$ peaks can already be observed for the semi-cubic superballs, which indicates that their chains are able to slide more freely with respect to each other than the spherical superballs, showing that a continuous transition from the hexagonal lattice to a full sliding phase occurs. Clearly, in the superball monolayers, the order in the direction of the mechanical rubbing remains due to chain formation, but the order perpendicular to the direction of the mechanical rubbing is quickly lost as the chains can slide along each other. These observations confirm the lattice structures seen in the SEM images shown in Figure 1d–f.

5. Conclusions

Monolayers of differently shaped silica superballs, namely spherical, semi-cubic and cube-like, were prepared using the unidirectional rubbing method and analyzed with SAXS.

In the 2D SAXS patterns some of the Bragg peaks appeared to be missing due to the presence of minima in the particle form factors. Due to the anisotropic shape and the hollow nature of the particles, it was difficult to predict which peaks are missing, which complicated the analysis of the patterns. By imaging the diffraction patterns of the monolayers over a range of rotations around the x-axis, the missing Bragg peaks were visualized. By doing so, the structural analysis was possible for all three superball shapes. We found that densely packed monolayer structures were obtained for all three types of particles using the unidirectional rubbing method. The spherical superballs formed hexagonal lattices with a high degree of positional order in all directions, whereas upon increasing the shape parameter m it was observed that this high degree of positional order in the direction of rubbing remained, but weakened in the direction perpendicular to the direction of rubbing, resulting in the formation of the sliding phase for the cube-like superballs. In the future, we will use this analysis method to further study and characterize the lattices of colloidal superballs assembled using the unidirectional rubbing method.

Thus, by exploring the three-dimensional reciprocal space of a two-dimensional sample we were able to effectively exclude the influence of the anisotropic form factor and analyze the densely packed structures of the colloidal monolayers in greater detail. As a final remark, we note that the problems discussed in the present work are not X-ray specific, but can as well be of importance for small-angle neutron scattering studies. If samples possess periodic ordering in fewer dimensions, in 1D or 2D, a rotation of the sample may allow the revelation of all essential diffraction features even in the case of Bragg peaks coinciding with form factor minima.

Author Contributions: Conceptualization, D.N.t.N. and A.V.P.; methodology, D.N.t.N. and A.V.P.; software, D.N.t.N.; validation, D.N.t.N., J.-M.M. and A.V.P.; formal analysis, D.N.t.N.; investigation, D.N.t.N. and A.V.P.; writing—original draft preparation, D.N.t.N.; writing—review and editing, D.N.t.N., J.-M.M. and A.V.P.; visualization, D.N.t.N.; supervision, J.-M.M. and A.V.P.; project administration, A.V.P.; funding acquisition, A.V.P. All authors have read and agreed to the published version of the manuscript.

Funding: This research was funded by NWO grant number 712.015.003 (D.N.t.N), 016.Veni.192.119 (J.-M.M.) and 195.068.1113 (ESRF beamtime).

Institutional Review Board Statement: Not applicable.

Informed Consent Statement: Not applicable.

Data Availability Statement: The data presented in the study are available on request from the corresponding author.

Acknowledgments: Dominique Thies-Weesie is thanked for her help with the preparation of the spherical silica colloids. Nicholas Orr is thanked for his help at the beamline of the ESRF. Michela Brunelli and Daniel Hermida Merino are thanked for providing technical support at the beamline. Cedric Gommes is thanked for useful discussions. The Nederlandse Organisatie voor Wetenschappelijk Onderzoek (NWO) is thanked for providing the financial support.

Conflicts of Interest: The authors declare no conflict of interest.

References

1. Boles, M.A.; Engel, M.; Talapin, D.V. Self-assembly of colloidal nanocrystals: From intricate structures to functional materials. *Chem. Rev.* **2016**, *116*, 11220–11289. [CrossRef] [PubMed]
2. Xu, Z.; Wang, L.; Fang, F.; Fu, Y.; Yin, Z. A review on colloidal self-assembly and their applications. *Curr. Nanosci.* **2016**, *12*, 725–746. [CrossRef]
3. Chen, Q.; Bae, S.C.; Granick, S. Directed self-assembly of a colloidal kagome lattice. *Nature* **2011**, *469*, 381–384. [CrossRef] [PubMed]
4. Yan, J.; Han, M.; Zhang, J.; Xu, C.; Luijten, E.; Granick, S. Reconfiguring active particles by electrostatic imbalance. *Nat. Mater.* **2016**, *15*, 1095–1099. [CrossRef]
5. Zhang, J.; Grzybowski, B.A.; Granick, S. Janus particle synthesis, assembly, and application. *Langmuir* **2017**, *33*, 6964–6977. [CrossRef]

6. Park, C.; Lee, T.; Xia, Y.; Shin, T.J.; Myoung, J.; Jeong, U. Quick, large-area assembly of a single-crystal monolayer of spherical particles by unidirectional rubbing. *Adv. Mater.* **2014**, *26*, 4633–4638. [CrossRef]
7. Park, C.; Koh, K.; Jeong, U. Structural Color Painting by Rubbing Particle Powder. *Sci. Rep.* **2015**, *5*, 1–5. [CrossRef]
8. Cademartiri, L.; Bishop, K.J.; Snyder, P.W.; Ozin, G.A. Using shape for self-assembly. *Philos. Trans. R. Soc. A Math. Phys. Eng. Sci.* **2012**, *370*, 2824–2847. [CrossRef]
9. Rossi, L.; Sacanna, S.; Irvine, W.T.; Chaikin, P.M.; Pine, D.J.; Philipse, A.P. Cubic crystals from cubic colloids. *Soft Matter* **2011**, *7*, 4139–4142. [CrossRef]
10. Meijer, J.M.; Byelov, D.V.; Rossi, L.; Snigirev, A.; Snigireva, I.; Philipse, A.P.; Petukhov, A.V. Self-assembly of colloidal hematite cubes: A microradian X-ray diffraction exploration of sedimentary crystals. *Soft Matter* **2013**, *9*, 10729–10738. [CrossRef]
11. Meijer, J.M.; Pal, A.; Ouhajji, S.; Lekkerkerker, H.N.; Philipse, A.P.; Petukhov, A.V. Observation of solid-solid transitions in 3D crystals of colloidal superballs. *Nat. Commun.* **2017**, *8*, 1–8. [CrossRef]
12. Meijer, J.M.; Meester, V.; Hagemans, F.; Lekkerkerker, H.N.; Philipse, A.P.; Petukhov, A.V. Convectively assembled monolayers of colloidal cubes: Evidence of optimal packings. *Langmuir* **2019**, *35*, 4946–4955. [CrossRef]
13. Meijer, J.M.; Hagemans, F.; Rossi, L.; Byelov, D.V.; Castillo, S.I.; Snigirev, A.; Snigireva, I.; Philipse, A.P.; Petukhov, A.V. Self-assembly of colloidal cubes via vertical deposition. *Langmuir* **2012**, *28*, 7631–7638. [CrossRef]
14. Rossi, L.; Soni, V.; Ashton, D.J.; Pine, D.J.; Philipse, A.P.; Chaikin, P.M.; Dijkstra, M.; Sacanna, S.; Irvine, W.T. Shape-sensitive crystallization in colloidal superball fluids. *Proc. Natl. Acad. Sci. USA* **2015**, *112*, 5286–5290. [CrossRef]
15. Bartlett, P.; Ottewill, R.H. A neutron scattering study of the structure of a bimodal colloidal crystal. *J. Chem. Phys.* **1992**, *96*, 3306–3318. [CrossRef]
16. Vos, W.L.; Megens, M.; Van Kats, C.M.; Bösecke, P. X-ray diffraction of photonic colloidal single crystals. *Langmuir* **1997**, *13*, 6004–6008. [CrossRef]
17. Scotti, A.; Gasser, U.; Herman, E.S.; Pelaez-Fernandez, M.; Han, J.; Menzel, A.; Lyon, L.A.; Fernández-Nieves, A. The role of ions in the self-healing behavior of soft particle suspensions. *Proc. Natl. Acad. Sci. USA* **2016**, *113*, 5576–5581. [CrossRef]
18. Otsuki, A.; de Campo, L.; Garvey, C.; Rehm, C. H2O/D2O Contrast Variation for Ultra-Small-Angle Neutron Scattering to Minimize Multiple Scattering Effects of Colloidal Particle Suspensions. *Colloids Interfaces* **2018**, *2*, 37.10.3390/colloids2030037. [CrossRef]
19. Klokkenburg, M.; Erné, B.H.; Wiedenmann, A.; Petukhov, A.V.; Philipse, A.P. Dipolar structures in magnetite ferrofluids studied with small-angle neutron scattering with and without applied magnetic field. *Phys. Rev. E Stat. Nonlinear Soft Matter Phys.* **2007**, *75*, 1–9. [CrossRef]
20. Krouglov, T.; Bouwman, W.G.; Plomp, J.; Rekveldt, M.T.; Vroege, G.J.; Petukhov, A.V.; Thies-Weesie, D.M. Structural transitions of hard-sphere colloids studied by spin-echo small-angle neutron scattering. *J. Appl. Crystallogr.* **2003**, *36*, 1417–1423. [CrossRef]
21. Holderer, O.; Zolnierczuk, P.; Pasini, S.; Stingaciu, L.; Monkenbusch, M. A better view through new glasses: Developments at the Jülich neutron spin echo spectrometers. *Phys. B Condens. Matter* **2019**, *562*, 9–12. [CrossRef]
22. Gommes, C.J.; Reiner, Z.; Jaksch, S.; Frielinghaus, H.; Holderer, O. Inelastic Neutron Scattering Analysis with Time-Dependent Gaussian-Field Models. 2021, to be published.
23. Riekel, C.; Burghammer, M.; Davies, R. Progress in micro- and nano-diffraction at the ESRF ID13 beamline. *IOP Conf. Ser. Mater. Sci. Eng.* **2010**, *14*, 012013. [CrossRef]
24. Petukhov, A.V.; Meijer, J.M.; Vroege, G.J. Particle shape effects in colloidal crystals and colloidal liquid crystals: Small-angle X-ray scattering studies with microradian resolution. *Curr. Opin. Colloid Interface Sci.* **2015**, *20*, 272–281. [CrossRef]
25. Stöber, W.; Fink, A.; Bohn, E. Controlled growth of monodisperse silica spheres in the micron size range. *J. Colloid Interface Sci.* **1968**, *26*, 62–69. [CrossRef]
26. Sugimoto, T.; Sakata, T. Preparation of monodisperse pseudocubic α-Fe$_2$O$_3$ particles from condensed ferric hydroxide gel. *J. Colloid Interface Sci.* **1992**, *152*, 587–590. [CrossRef]
27. Snigirev, A.; Kohn, V.; Snigireva, I.; Lengeler, B. A compound refractive lens for focusing high-energy X-rays. *Nature* **1996**, *384*, 49–51. [CrossRef]
28. Jiao, Y.; Stillinger, F.H.; Torquato, S. Optimal packings of superdisks and the role of symmetry. *Phys. Rev. Lett.* **2008**, *100*, 2–5. [CrossRef]

MDPI
St. Alban-Anlage 66
4052 Basel
Switzerland
Tel. +41 61 683 77 34
Fax +41 61 302 89 18
www.mdpi.com

Applied Sciences Editorial Office
E-mail: applsci@mdpi.com
www.mdpi.com/journal/applsci

www.ingramcontent.com/pod-product-compliance
Lightning Source LLC
LaVergne TN
LVHW070657100526
838202LV00013B/982